6th Target Fabrication Workshop (TFW6) and 3rd Targetry for High Repetition Rate Laser-Driven Sources Conference (Targ3)

Journal of Physics: Conference Series Volume 1079

London, United Kingdom
8-11 May 2017
Salamanca, Spain
21-23 June 2017

Editors:

Christopher Spindloe

ISBN: 978-1-5108-7088-8
ISSN: 1742-6588

Printed from e-media with permission by:

Curran Associates, Inc.
57 Morehouse Lane
Red Hook, NY 12571

CURRAN ASSOCIATES INC. proceedings .com

Some format issues inherent in the e-media version may also appear in this print version.

Printed by Curran Associates, Inc. (2018)

Institute of Physics
Dirac House, Temple Back
Bristol BS1 6BE UK

Phone: 44 1 17 929 7481
Fax: 44 1 17 920 0979

techtracking@iop.org

Additional copies of this publication are available from:

Curran Associates, Inc.
57 Morehouse Lane
Red Hook, NY 12571 USA
Phone: 845-758-0400
Fax: 845-758-2633
Email: curran@proceedings.com
Web: www.proceedings.com

Table of contents

Volume 1079

2018

6th Target Fabrication Workshop (TFW6) and the Targetry for High Repetition Rate Laser-Driven Sources (Targ3) Conference
8–11 May 2017, University of Greenwich, London (TFW) and 21–23 June 2017 Salamanca, Spain (Targ)

Accepted papers received: 2 August 2018
Published online: 6 September 2018

Preface

Papers

Targ3

6TFW

IOP Conf. Series: Journal of Physics: Conf. Series **1079** (2018) 011001 doi:10.1088/1742-6596/1079/1/011001

Conference Proceedings, Editor's Preface

2017 was an exciting year for the Target Fabrication community with both the 6th Target Fabrication Workshop (TFW) held in Greenwich London, UK and the 3rd Targetry for High Repetition Rate Laser Driven Sources (Targ) workshop held Salamanca, Spain taking place.

These well-established workshops, with their focus on different but complimentary areas of Target Fabrication, cover the main areas of research providing novel and high specification samples for the next generation of high power laser systems. The TFW series, has developed to cover high repetition rate solid targets, cryogenic targets and microfluidics and had exciting talks from many PhD students from around the world. The TargN workshop series brings together experts and key players involved in the latest trends on intense laser-plasma research, requiring the design of specific targets aimed at particular requirements to exploit unique opportunities presented by such extreme peak laser powers.

As it is becoming apparent that experimental campaigns are no longer shot, but target limited these conferences are a valuable resource to the emerging laboratories, industrial collaborators and students to share information and build collaborations to meet the new challenges in the field.

The opportunity for many researchers to attend both conferences within a short period of time encouraged exciting discussions between groups. These discussions, across the whole range of Target Fabrication activities, and the opportunity to present both conference proceedings in a single journal was a useful record and resource for the community in the future.

On behalf of my co-editors I would like to thank all who have contributed to these Proceedings and we look forward to future conferences in this developing field.

Christopher Spindloe
Principal Editor

IOP Conf. Series: Journal of Physics: Conf. Series **1079** (2018) 011002 doi:10.1088/1742-6596/1079/1/011002

Third Targetry Meeting (Targ3), Salamanca, Spain

The third targetry meeting (Targ3) was organised from 21st to the 23rd of June 2017 at the Centro de Laser Pulsado (CLPU) in Salamanca, Spain and was focused on the latest targetry advances for laser-driven particle acceleration and laser radiation sources. The TargN workshop series brings together experts and key players involved in the latest trends on intense laser-plasma research, requiring design of specific targets aimed at particular requirements to exploit unique opportunities ushered in by such extreme peak laser powers. The three-day meeting attracted almost 50 researchers and students from different countries in Europe and abroad: UK, USA, JAPAN, Germany, Italy, France, Spain, Romania, Israel and Czech Republic.

The involved institutions are:

Institute of Optoelectronics, Military University of Technology; Centro de Laser Pulsado (CLPU); ELI Beamlines; European XFEL; Laboratoire d'Optique Appliquée; Ecole Polytechnique; Ludwig-Maximilians Universität, München; Instituto de Microelectronica de Barcelona, IMB-CNM-CSIC; Helmholtz-Zentrum Dresden-Rossendorf; STFC Rutherford Appleton Laboratory; Lawrence Livermore National Laboratoryl; Ohio State University; Max-Planck-Institut für Quantenoptik, University of Jena; Technical University Darmstadt; University of Salamanca; CEA Saclay; Politecnico di Milano; Universidad de Santiago de Compostela; Hebrew University of Jerusalem; ELI-NP.

Following in line with the first two editions (in October 2013 in Garching (Germany), and April 2015 in Paris (France)) the percentage of young scientist was maintained above 50% and their contributions, the levels of complexity and quality of their research was outstanding. The conference was organised keeping the number of participants below a defined number to not be dispersive and maximise the time of free discussion around the talks in order to facilitate the flowing of the information.

Targetry topics were discussed from experimental, fundamental and technological points of view. Target production, optimization and composition was presented, analysed and compared with the technological state of the art of laser and target areas. For example, sophisticated techniques for thin target production were compared with methods for laser contrast optimisation and the three different possible states of targets: solid liquid and gaseous were discussed in many talks comparing technological quality with the final efficiency in interaction with lasers.

The continuous interchange between scientific interests and the necessary technological developments represents the biggest challenge of the modern laser infrastructures and involves different curriculums in physics (Nuclear, Plasma, Laser ....) in engineering, chemistry and more. The presence of more than 50 % of young scientist is a good sign for future developments and collaborations in this area.

Luca Volpe
Chair of the local organising committee

Joerg Schreiber, François Sylla, Rodrigo Lopez-Martens.
TargN Committee.

IOP Conf. Series: Journal of Physics: Conf. Series **1079** (2018) 011002 doi:10.1088/1742-6596/1079/1/011002

IOP Conf. Series: Journal of Physics: Conf. Series **1079** (2018) 011003 doi:10.1088/1742-6596/1079/1/011003

6th Target Fabrication Workshop, Greenwich, London, UK

The 6th Target Fabrication Workshop was held at the University of Greenwich, London, 8 - 11 May 2017. Approximately 55 delegates took part from institutes in: UK, France, Germany, Spain, Hungary, Romania, US and Japan. Oral and poster presentations were given on a variety of Target Fabrication technologies used for manufacture of target for irradiation in high-energy laser experiments in institutes across the world. In depth discussions were held on collaborative opportunities between departments, with particular interest in creating an international delivery network to cater for all requirements, from any facility using the expertise from the wider Target Fabrication community.

Attendees included representatives from institutions and companies such as: Rutherford Appleton Laboratory(UK), University of Michigan(US), LMU (Germany), Lawrence Livermore National Laboratory(US), AWE(UK), LULI(France), CEA(France), University of Darmstadt(Germany), ELI-NP(Romania), ELI – ALPS(Hungary), IMPMC (France), UCL(UK), The Institute of Laser Engineering(Japan), Cardiff University(UK), Nottingham University (UK),Tokyo Institute of Technology(Japan), General Atomics(US), Politecnico di Milano(Italy), Universidad Politecnica Madrid(Spain), HZDR(Germany), Luxel Corporation(US), SourceLAB(France), Applied Diamond(US) and Scitech Precision Ltd(UK). There were as many similarities as differences between international institutes, some had astonishing throughput of a single complex target design with characterisation data to match. Other institutes and commercial entities were more geared towards a variety of experimental technologies with an eye to the next 10 years of target fabrication.

Talks and discussions took a workshop format; questions were held at the end of a three talk session where the speakers returned to the front, this encouraged a broader discussion on the topic and all speakers could offer their expertise simultaneously. At points discussions involved the whole delegation with institutes offering their resources to help solve any issues encountered.

The talks were split into six different sessions: facility overviews, Micro-target Fabrication, Cryogenics, High-repetition rate, Silicon based fabrication, and, Micro-fluidics and Chemistry-based techniques.

The previous workshop was held in University of St Andrews, Scotland two years prior and it was exciting to see how far development has progressed. For example, at TFW5 there were presentations focused on early development of high-repetition rate solutions that have now been experimentally fielded, developments in cryogenic targetry where systems in their infancy at TFW5, had been fielded on multiple campaigns, and many other presentations able to reflect on how far technology had progressed.

Micro-fluidics has also come of age since TFW5, delegates from Cardiff University were able to offer concluding remarks on three years of PhD work and offered an opportunity for delivery to experiments in the near future.

A keynote presentation was given by a theoretical modelling representative, Alex Robinson, giving a perspective on how target design and quality can affect interactions in high-power laser experiments.

IOP Conf. Series: Journal of Physics: Conf. Series **1079** (2018) 011003 doi:10.1088/1742-6596/1079/1/011003

Without the financial Sponsorship provided by: AWE, General Atomics, SourceLAB, Scitech Precision and TUD the workshop would not have been possible. On behalf of the committee and delegates we would like to thank all of them.

The next target fabrication Workshop has been scheduled and the organising committee look forward to welcoming the community to Darmstadt, Germany in September 2018.

David Haddock
Chair of the local organising committee

Martin Tolley, Christopher Spindloe, Gabriel Schaumann, Wigen Nazarov
TFW Committee

IOP Conf. Series: Journal of Physics: Conf. Series **1079** (2018) 011004 doi:10.1088/1742-6596/1079/1/011004

Peer review statement

All papers published in this volume of *Journal of Physics: Conference Series* have been peer reviewed through processes administered by the proceedings Editors. Reviews were conducted by expert referees to the professional and scientific standards expected of a proceedings journal published by IOP Publishing.

IOP Conf. Series: Journal of Physics: Conf. Series **1079** (2018) 012001 doi:10.1088/1742-6596/1079/1/012001

Experimental study of proton acceleration from thin-foil on a table top Ti:Sapphire

I Sánchez, R Lera, J Ruda, JB Gonzalez, F Ruiz, D Lopez, A Ruiz-de la Cruz

Proton Laser Application S.L. (PLA), Av. Vilafranca del Penedés 11ª, 08734 (Barcelona), Spain.
E-mail: i.sanchez@protonlaser.com.

Abstract. Table-top lasers of moderate energies (0.1- 0.5 J) are promising, cost-effective laser-driven sources of MeV-energy protons and ions accelerators, but many applications of laser-accelerated ion beams will not only require a low cost, compact system but also require operation at high repetition rates (>10 Hz).

This paper focus the effort to characterize and optimize proton acceleration processes using 3 TW/55 fs, table-top Ti:Sapphire (Ti:Sa) laser capable to operate at 10 Hz, with this intention the maximum proton energy obtained, in a single shot, was evaluated by irradiating a variety of flat targets, from thin film to sub-micrometric film membranes designed for our purposes in a high density target array.

The presented work is a previous stage and this study has the intention to be the prelude of the technological challenge, both in laser technology as well target design capable of operating at high repetition rates.

1. Introduction

Over recent years considerable effort has been made to develop laser-driven ion sources for multiple applications ranging from ion oncology [1] to proton radiography [2, 3, 4, 5]. Moreover, many applications of laser-accelerated ion beams will not only require a low cost, compact system, but also require operation at high repetition rates to achieve the necessary particle flux. e.g., radiation therapy [3], are demanding average flux which requires an elevated shot rate to accumulate useful levels of activity [6].

High intensity Ti:Sa lasers sources demonstrates that they not only occupy a smaller spatial, but they are able to operate at significantly higher repetition rates when compared to similar intensity Nd:Glass lasers [6,7]. Ti:Sa lasers are most likely to fulfill these requirements, hence experimental effort to characterize and optimize ion acceleration processes using femtosecond laser Ti:Sa sources is a crucial step towards many proof of principle experiments [8, 9].

Proton or ion acceleration with 10 Hz systems has been demonstrated at several laboratories [10, 11]. However, the actual shot rate normally is much lower due to restrictions in the target positioning which typically requires a mechanical precision of a few microns.

This is a real challenge that has not yet been solved and where we wanted to focus our efforts, in that sense this work presents an experimental study of laser-driven ion acceleration with micrometer and submicrometer thick Al flat targets, the fabrication of these targets is another technological challenge because the repetition rate is determined by two factors, the pulse rate of the laser system and the time necessary to refresh the target [10-16].

The subjects to the present work are the follows: Firstly, we can point out the development of a compact table top Ti:Sa, low-cost, energy efficient, stable, ultra-high-intensity and ultra-short pulse laser with 10 Hz repetition rate. For this purpose, this paper reports the measurements of proton production using pulse duration of tens of femtoseconds ~55 fs and energies below 200 mJ. Optimum conditions for proton acceleration in terms of laser energy, contrast, target material and thickness have been characterized. Secondly, we employ large arrays of solid targets as the beginning for future high repetition processes, due to the complexity of the experiment here we show the single shot results, but the results obtained with our table-top laser could be demonstrated for accelerated protons and set high requirements on the laser and target properties.

2. Experimental setup

2.1.Laser system

Experiments was carried out employing a Chirped Pulse Amplification (CPA) Ti:Sa laser developed at Proton Laser Applications S.L (PLA) [12, 13]. The laser can deliver pulses energy up to 275 mJ (160 mJ on target surface) of energy with duration of 55 fs at a central wavelength of 780 nm. The scheme of the laser developed in this work is presented in Fig.1 and detailed in this section.

As the figure 1 shows, the front end consists of a commercial Ti:Sa oscillator (Venteon of Laser Quantum) that delivers 10-fs pulses with 80 MHz repetition frequency. Then the pulse passes through a stretcher which enlarges the pulse 10000 times. The stretcher employed was a single pass Öffner type [14] chosen due to its simplicity and its aberration low design [15], the mentioned stretcher comprised of one gold-coated reflecting diffraction grating (Horiba Jovin-Yvon) with a line density of 1200 lines/mm and 92 % reflectivity at 800 nm, tilted at 60° and a relay system of two spherical mirrors with a radius of curvature of 500 mm and -250 mm respectively. In order to compensate third order dispersion components due to the stretcher process it is required employ an external device, in our case it was used a programmable acousto-optic device (Dazzler of Fastlite) capable of shaping both the spectral intensity and phase of stretched pulse. At the end of the process the pulse duration of the stretched pulse was estimated to be more than 150 ps using a fast photodiode.

Afterwards, the remaining signal (6 nJ) is amplified first in a regenerative pre-amplifier which increase the energy to 1.5 mJ per pulse and achieve a beam diameter of 2 mm. As is shown in the figure, the active medium is a Ti:Sa crystal with rectangular end faces of 4 x 6 mm and 20 mm long. The pump laser of the regenerative amplifier is an Empower 15 of Spectra-Physics, which delivers 8 mJ and gaussian shaped beams of 200 ns at 100 Hz. Preceding the next amplifier, a Faraday isolator is included to prevent backreflected light from the following amplifiers.
Subsequently, a second three-pass pre-amplifier pumped at 532 nm by a Nd:YLF laser increased the energy to 30 mJ. This multipass amplifier is pumped by up to 200 mJ in 15 ns pulses at 100 Hz, this nanosecond laser deserves special mention because were designed to be used to pump this material and play a fundamental role to achieve a good performance at high repetition rates. It includes two diodes pumping chambers, PH5400 (manufactured in Monocrom S.L.) [12], loaded with Nd:YLF rods of 5 mm in diameter, there are disposed as a Master Oscillator Power Amplifier (MOPA) configuration. A KTP

IOP Conf. Series: Journal of Physics: Conf. Series **1079** (2018) 012001 doi:10.1088/1742-6596/1079/1/012001

crystal converted to a second harmonic the output energy at 527 nm and pump a three multipass configuration which contains a cylindrical Ti:Sa crystal 6 mm diameter x 15 mm length, water cooling. In order to compensate the action of the thermal lensing in this amplifier and improve the extraction efficiency, we implemented a telescope situated prior to the first pass which introduces divergence and compensate the converging effect induced by thermal lensing. The beam diameter obtained is 2.5 mm with 30 mJ of pulse energy.

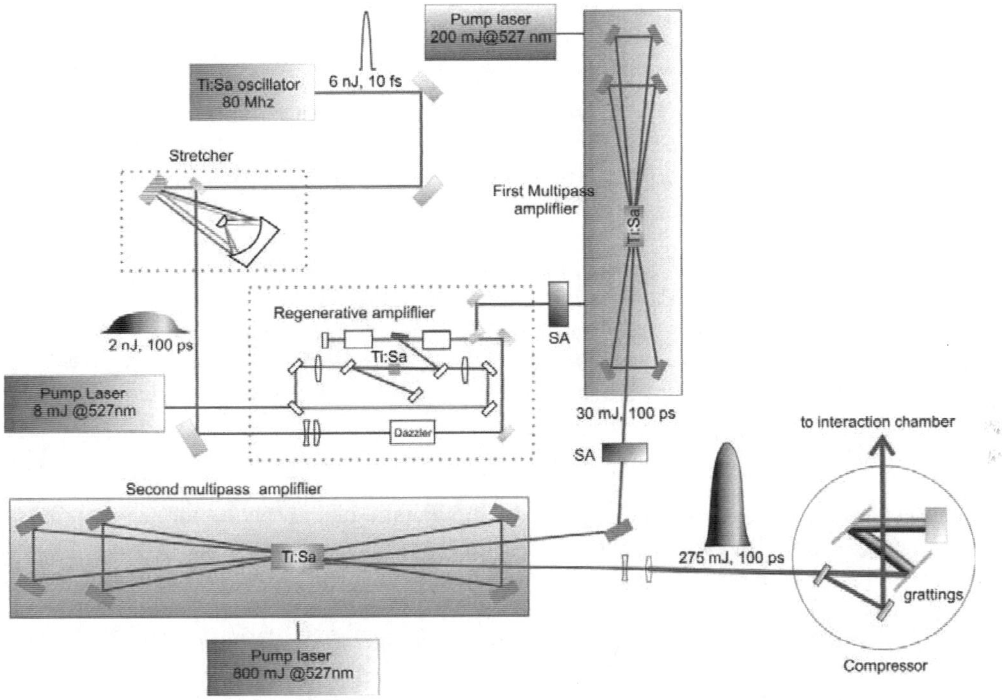

Figure 1. a) Scheme of the table-top Ti:Sa laser developed at Proton Laser Applications S.L (PLA). A pulse oscillator sends a beam of a few femtoseconds that is stretched in time in a pulse stretcher. Then it enters a chain of amplification that starts with a regenerative amplifier that is followed by a first multipass amplifier and ends with a second multipass amplifier. The CPA chain ends with a compressor installed in a vacuum chamber.

After the first multipass amplifier the laser pulse is introduced into the final three-pass amplifier which is pumped with two 400 mJ, 532-nm Nd:YLF lasers and reaches uncompressed 275-mJ energy level. The amplifier is built in a multipass bow tie configuration with 3 passes. The Ti:Sa active medium is a plane-parallel crystal with 16 x 20 mm^2 rectangular facets and 20 mm length. The Crystal is cooled in cooper mount water cooling, which permits refrigerate the crystal at 15°C, to mitigate the effect of the thermal lensing, several approaches were used. First the seed beam was enlarged thanks to a telescope of low magnification introducing divergence, additionally, a convex mirror of radius of curvature -2000 mm had to be inserted before the last pass to increase the divergence of the laser and to prevent the beam from leaving the amplifier converging. The output beam had a diameter of 5.3 mm and 275 mJ energy per pulse are achieved using a total pump of 800 mJ.

In order to clean the pulse two saturable absorbers (SA) were inserted after the regenerative amplifier and the second before the final multi-pass amplifier. The contrast ratio of the amplified spontaneous (ASE) at a few ns was estimated to be 10^{-8} prior to the peak of the main pulse, and 10^{-6} of the peak pulse intensity at a few ps. These measurements were made by using a fast photodiode and oscilloscope. In ultra-intense lasers, the temporal quality of the pulses is of maximum importance, especially in applications such as particle acceleration. TW laser can be focused to a few microns and achieve a peak intensity more than 10^{18} W/cm^2, if the contrast at some point is less than six orders of magnitude, a preceding part of the pulse would present an intensity of at least 10^{12} W/cm^2, which can be enough to interact non-linearly with a material and generate plasma [10, 15].

Finally, the amplified pulse is guided to a compressor (63 % efficiency) with a standard configuration based on Treacy's design [15] and covered inside a vacuum chamber. Consider the damage threshold of the holographic grating is about 300 mJ/cm^2, the beam was enlarged by a series of mirrors and a telescope to a diameter of 20 mm.

The pulse duration was measured with a Wizzler (Fastlite), which uses the third-order technique of self-referenced spectral interferometry [16]. A feedback loop can be set up between this device and the spectral shaper located before the regenerative amplifier to minimize the dispersion, especially third order, which was found to be the term which contributed the most to aberrations in the spectral phase. The pulse duration was determined to be 55 fs and the spectral bandwidth retrieved was 19 nm FWHM.

2.2.Interaction Chamber

The proton acceleration experiments take place in a vacuum chamber (20 mtorr) situated after the compressor (Fig. 2). A pulse coming from the compressor is p-polarized and is focused on a target at an incident angle of 30 ° by using an f/1.3 off-axis parabola mirror (OAP). The spot diameter was measured with a CCD (WinCamD-UCD12 of Dataray) camera with an aspheric lens 0.57 µm effective pixel size, where we obtained the full width-half-maximum of the focal size of 5 × 10 µm^2. The parameters, which have routinely been achieved on the target for experiments, are listed in Table 1.

Table 1. Characteristic Ti:Sa laser parameters used in this study.

Parameter	Variable unit	Value
Pulse duration	τ_{FWHM} (fs)	55
Focal spot size	D_{FWHM} (µm)	5 × 10
Energy on sample	E_L (J)	0.160
Power on sample	P_L (TW)	3
Contrast		10^{-8}

Focus alignment, which has a sensitivity of few micrometers, is controlled through a He:Ne laser coupled on the same optical path of the Ti:Sa beam [12, 13], a CCD camera imaging the systems, is placing along optical path of Ti:Sa laser beam and serves the dual purpose of laser focus diagnosis and precise positioning of the target.

The target holder formed by a framework with 16 windows to hold up the targets, these windows has a specific profile for tension the surface of the target and permits avoid variations from the exact focusing position arising from target roughness or flatness. The holder, positioned at 30° of the incident beam, is moved by three motors enabling three freedom degrees with a positioning precision of 2.5 um. This configuration allows for testing several targets without breaking the vacuum condition.

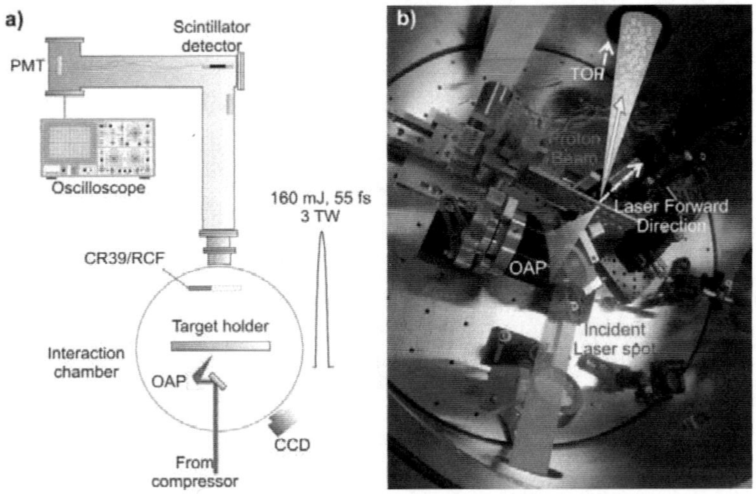

Figure 2. a) Pictured in the image a schematic view of the interaction chamber and the disposition of the detectors employed in this experiment, behind the target holder there is a passive radiation detector (CR-39), and at the end of the flight tube there is a scintillator detector coupled by fiber to a photomultiplier tube (PMT). b) detailed view of the interaction chamber. In the first vacuum chamber a mirror steers the laser beam onto the OAP (off-axis parabola), which focuses the laser onto a target where the protons are generated.

According to the established theoretical models [18-20], the process acts as in the following: a high intensity laser impinges on the front surface of a thin foil, suddenly some electrons are directly accelerated by and penetrate the target. Most them spread and dissipates energy inside of it while the hot component of these electrons can reach the target rear side [20, 21]. Only the most energetic of these electrons can escape, leaving behind an electrostatic potential which generates an electric field that ionizes and accelerates surface ions in a process called Target Normal Sheath Acceleration (TNSA) [19-25].

2.3.Sample description

As mentioned previously, one of the bottleneck for a real laser-driven ion application are the rate samples supply system matched to the repetition rate of the driving laser. This is a field of technological research that is attracting a lot of attention, in our case we make a parallel research in collaboration with the Institute of Microelectronics of Barcelona (IMB-CNM (CSIC)), where a target array for high repetition rate, was designed [26].

For comparative reason, the experimental campaign was made with series (*Serie 1*) of different thickness Aluminum (Al) foils. This campaign allows us to find the optimal laser parameters in laser driven proton acceleration. A second campaign was made with a fabrication of sub-micrometric thin-layer conductive membranes embedded in a silicon wafer frame by using nano/micro-electro-mechanical-system

(N/MEMS), these samples are made by aluminum membranes supported by SiO_2 and are referred as S*erie 2*. A schematic structure of the samples is detailed in Figure 3.

Table 2. Sample description employed in this work, the samples are divided into two categories by composition (*Serie 1 or 2*), in which different thicknesses have been used.

Samples	Composition	Thickness *d (μm)*
Serie 1	Flat Al	0.8, 1.8, 2.5, 7.0
Serie 2	Flat Al +SiO_2	0.25, 0.5, 0.65, 1.0

The second method of fabrication of *series 2* targets, not only permit us to tune membranes properties, such as thickness and material composition, further, it is potentially applicable on high repetition rate regime because this target array can load up in each 16 individual holder windows permitting a total of 256 targets available for experiments within one vacuum cycle. This work is focused on the laser development description for laser acceleration experiments, the target fabrication deserves a separate chapter which is detailed in the following references [26, 27].

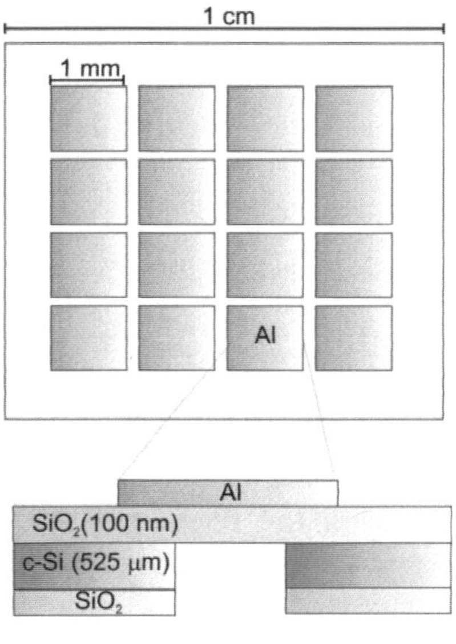

Figure 3. Structure of the samples, *Series 2*, formed by micrometric thin-layer conductive membranes embedded in a silicon wafer and amplified view of these test structure cross section.

2.4. Measurement and characterization techniques

The main diagnostic of multi-MeV proton acceleration was made by passive CR-39 nuclear track detectors and time of flight (TOF) measurements (see Fig 2).
Inside the same vacuum chamber, passive detectors such as 1 cm^2 CR-39 (Radosys) chips was placed in the direction behind the target. CR-39, allyl diglycol carbonate, is a plastic polymer, when an energetic proton arrives at the chip, it destroys the links between polymer chains through its range of penetration. However, these damaged tracks are not visible and the CR-39 must be post-treated with an etching process

with a bath of 6.25M NaOH at 90°C for 4 hours. The energy of the particles can be inferred from the shape and size of the pits.
The other measurement technique was a Time-of-Flight (TOF) configuration formed by scintillator crystal. This device was situated at the end of the time of flight tube with a distance of 227 cm from the interaction chamber. The scintillator detector was coupled to a photomultiplier tube (PMT) which generates a signal that can be recorded with a high bandwidth oscilloscope. The number of ions and therefore the absolute proton spectral distribution can be inferred from the height of the signal created by the PMT which was previously calibrated, together with the CR-39, at the 3 MV cyclotron at the National Accelerator Center (CNA), Seville [20]. The particle detection adapted to our system was developed and calibrated by the research Institute for Molecular Imaging (I3M, Valencia), more information of the calibration work is reported in references [13, 28]

From these techniques is possible to obtain complete spectra of number of protons as a function of the accelerated energy. Moreover, TOF measurement technique gives on-line real-time information about the ion current shot-to-shot fluctuations as well as the maximum proton energy in real time, due to its simple and fast readout, can be used as a real-time diagnostic with high repetition rate lasers.

3. Results and Discussion

3.1. Target thickness and contrast influence

An intense collimated beam of high-energy protons is emitted normal to the rear surface of thin solid targets irradiated at 3 TW power, the beam profile at the focus (5 x 10 μm^2 FWHM) was measured directly with a camera. Up to 160 mJ of the laser is transferred to the sample with a peak intensity is 2 $\times 10^{18}$ W/cm^2.

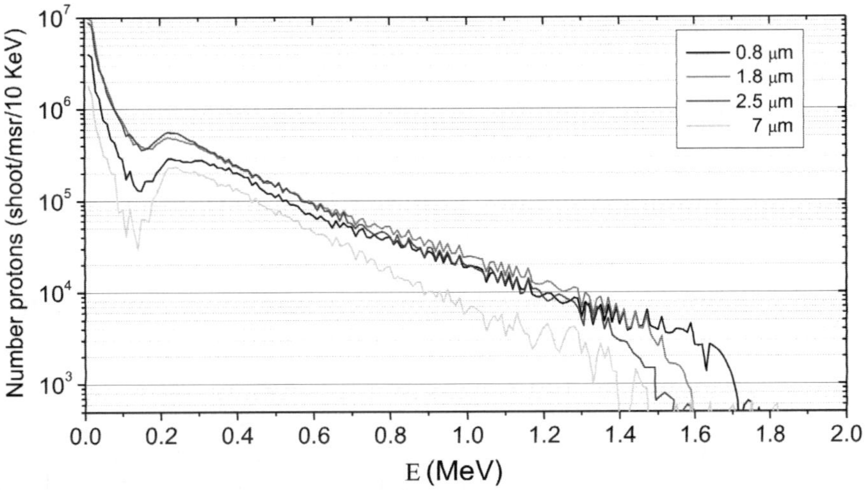

Figure 4. Typical proton density distribution from time of flight measurements comparing different results varying the Al target thickness. The plot shows de number proton density as a function of accelerated energy for different target thickness.0.8 μm (1.75 MeV), 1.8 μm (1.6 MeV), 2.5 μm (1.5 MeV), 7 μm (1.4 MeV).

IOP Conf. Series: Journal of Physics: Conf. Series **1079** (2018) 012001 doi:10.1088/1742-6596/1079/1/012001

In order to observe the distribution of the protons generated in the interaction of the laser and the target the TOF and the scintillator were used as the principal detector simultaneously CR-39 chips were used as a secondary source of information.

The spectrum of the accelerated protons measured by the TOF for targets with different thicknesses is presented in Fig. 4. The energy spectra of the proton beams observed, typically with an exponential profile, has three main features: a large energy spread, a smooth decrease of particle number with energy and finally an abrupt "cutoff," at a well-defined energy which we will term "maximum proton energy" in the following. This maximum proton energy is the quantity normally used to compare different experiments and determine experimental scaling laws for the acceleration process.

From these results we can obtain the maximum proton energy as a function of the target thickness. In this specific case, the maximum energy obtained with 0.8 μm reach 1.75 MeV, in the same graph we show the proton spectra obtained from 1.8 μm, 2.5 μm, and 7 μm Al thick.

With the objective to find the mechanism that can improve the maximum energy of the protons accelerated, it was measured the maximum proton energy for a variation of the thickness of the aluminum foil. Figure 5 shows the results achieved with different laser contrast. All the error bars were largely determined by the variation in maximum energy over multiple repeat data shots for each target thickness.

In that point we have to remark the importance of the contrast in the experiments, previous results in which we had a low contrast ($< 10^{-6}$) provided proton energies lower than with high contrast (10^{-8}) after introducing the saturable absorbers, the inclusion of these absorbers reduce the effect of the prepulse allowing to reach higher maximum energies and permitting the use of thinner targets, therefore it is evidence that for acceleration efficiency, contrast was an important parameter.

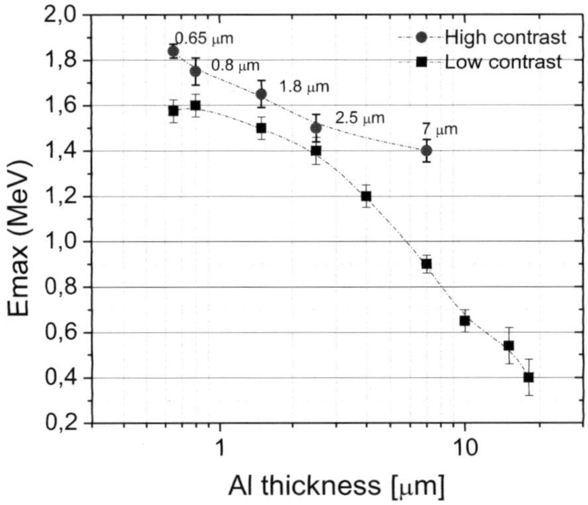

Figure 5. Maximum proton energy as a function of the flat Al target thickness, plot of measured data, color dots in blue correspond to high contrast experiments and black for low contrast conditions.

IOP Conf. Series: Journal of Physics: Conf. Series **1079** (2018) 012001 doi:10.1088/1742-6596/1079/1/012001

From the results obtained it has been shown that the energy maximum of proton accelerated via TNSA can be enhanced by using thinner targets and the proton acceleration depends strongly on the contrast of the laser. The comparison of our results with other experiments in scientist literature demonstrates that the proton energies observed in our experiments are consistent with theoretical scaling laws and previous empirical results [24, 25].

3.2. Proton acceleration with arrays of solid targets

Once we have the first campaign of proton acceleration, based on maximum proton energy from thin Al flat foil. Next step was to obtain the optimal laser-driven proton acceleration from the target *Serie 2*, based on MEMS fabrication, one of the scopes is to demonstrate the performance of targetry able to work at high repetition in vacuum.

In this stage same laser conditions described previously was performed using targets of *Serie 2*. The fabrication of these targets permits us to reduce the Al thicknesses until to 0.25 μm. Figure 6 shows the maximum proton energy as a function of Al thicknesses, comparing the different targets employed in this study, both *Serie 1* and *Serie 2*.

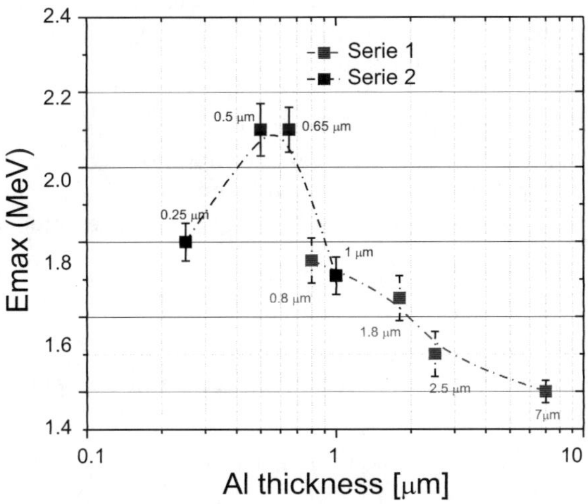

Figure 6. Maximum proton energy as a function of the flat Al target thicknesses.

From the previous plot it is shown the increase of maximum proton acceleration with targets of *series 2*, hence it is clear that the deigned samples at *Serie 2* match better with our laser system requirements and the results of preliminary experiments had shown the achievement of successful proton acceleration up to 2 MeV. Although this energy value is far from that necessary for the feasibility of laser-based proton accelerators, it is a very promising result and give us the importance of adapt the target composition and thicknesses to any specific laser-driven processes.

Another relevant question, on the contrary, we expected to reach the maximum proton energy with 0.25 μm thickness but the optimum value is achieved with 0.5 μm of Al thicknesses. How the acceleration varies in *series 2,* is still an open question and further investigation, including membrane composition,

IOP Conf. Series: Journal of Physics: Conf. Series **1079** (2018) 012001 doi:10.1088/1742-6596/1079/1/012001

focusing and laser parameters should be considered to fully understand these results. However, this study also demonstrates the existence of an optimum foil thickness and target design for reach maximum proton acceleration. Moreover, it has been shown that novel targets offer the possibility of improving the maximum energy and the number of individual targets per area have become manifest.

4. Conclusions

In this article we investigate laser-driven ion acceleration with micrometer and submicrometer Al flat targets and ultrashort 55 fs laser pulses with moderate energy (0.16 J) obtained from a Ti:Sa laser system. The main importance of our work resides in the successful operation of a compact, cost-effective laser ion accelerator. In that sense, the data obtained from planar foil targets demonstrate the effectiveness in proton acceleration with a table-top TW-class fs laser system capable of working at 10 Hz.

The improvement of contrast is an important contribution to the increase in proton energy since it enabled the use of thinner targets. The concept of contrast has low impact on the performance of the laser but can be very important to the efficiency of certain applications.

Also, we pay attention in the target and membranes properties, such as thickness for potential laser-plasma applications based on high repetition rate regime. The future research line is focused, from one hand to increase the laser pulse energy, and from the other, to develop more sophisticated target designs should improve the coupling and thus the acceleration results.

Laser-based proton accelerators are still at an early stage of evolution and major developments are necessary, hence experimental effort to characterize and optimize ion acceleration processes is a fundamental step towards achieving this objective and our system can be a valuable tool for studies the underlying physics of laser-plasma interactions as well as technical developments related to laser-based proton beams.

5. References

[1] Borghesi M, (2014). Laser-driven ion acceleration: State of the art and emerging mechanisms. Nucl Inst A, Vol. 740, 11.03.2014, pp.6–9.

[2] Green J.S, Robinson A. P. L, Booth N, Carroll D. C., Dance R. J., Gray R. J., MacLellan D. A, McKenna P, Murphy C. D., Rusby D, & Wilson L, (2014). *High efficiency proton beam generation through target thickness control in femtosecond laser-plasma interactions*. Appl. Phys. Lett. 104, 214101.

[3] Schardt, D. (2007) Heavy-Ion Therapy Collaboration. Tumor therapy with high-energy carbon ion beams. Nucl. Phys. A 787, 633.

[4] Schreiber, J., Bolton, P.R. & Parodi, K., (2016)."hands-on" laser-driven ion acceleration: A primer for laser-driven source development and potential applications. Rev Sci Ins, 87, 071101.

[5] Fourmaux, S. et al., (2013). Investigation of laser-driven proton acceleration using ultra-short, ultra-intense laser pulses. Physics of Plasmas, 20, 013110.

[6] Malka, V. Fritzler S, Lefebvre E, d'Humières E, Ferrand R, Grillon G, Albaret C, Meyroneinc S, Chambaret JP, Antonetti A, Hulin D. (2004) *Practicability of proton therapy using compact laser systems*. Med. Phys. 31(6): 1587-92.

IOP Conf. Series: Journal of Physics: Conf. Series **1079** (2018) 012001 doi:10.1088/1742-6596/1079/1/012001

[7] Safronov, K. V. et al., (2009). Experimental study of the acceleration of protons emitted from thin foils irradiated by ultrahigh-contrast laser pulses. JETP Letters, 88(11), pp.716–719.

[8] Spencer I et al. (2003) Experimental study of proton emission from 60-fs, 200-mJ high-repetition-rate tabletop-laser pulses interacting with solid targets. Phys, Rev, E 67 046402.

[9] Rosinski, M. et al., (2016). Acceleration of protons in plasma produced from a thin plastic or aluminum target by a femtosecond laser. Journal of Instrumentation Vol 11, SN - 1748-0221N2.

[10] Daido H, Nishiuchi M, & Pirozhkov A.S, (2012). *Review of laser-driven ion sources and their applications. Reports on progress in physics.* Physical Society (Great Britain), 75(5), p.56401.

[11] T. Ceccotti, A. L_evy, H. Popescu, F. R_eau, P. D'Oliveira, P. Monot, J. P. Geindre, E. Lefebvre, P. Martin, Proton acceleration with high-intensity ultrahigh-contrast laser pulses, Phys. Rev. Lett. 99 (2007) 185002.

[12] Matellanes, R. L. (2017). Thesis : Design and development of a high power high repetition rate ultrashort pulse laser for proton acceleration. University of Salamanca, Spain.

[13] Bellido P, Lera R, Seimetz M, Zaffino R. L, Vidal L, Soriano A, Rigla J. P, Moliner L, Iborra A, Conde P, Aguilar A, and Roso L, (2016) *Characterization of protons accelerated from a 3 TW table-top laser system*, JINST, 12 p. T05001, 2017.

[14] A. Öffner, U.S. patent 3,748,015 (1971).

[15] G. Cheriaux, B. Walker, L. F. Dimauro, P. Rousseau, F. Salin, and J. P. Chambaret, *Aberration-free stretcher design for ultrashort-pulse amplication*, Opt. Lett. 21, 414{416 (1996).

[16] E. B. Treacy, *Optical pulse compression with diffraction gratings*, IEEE J. Quantum Electron. 5 (9), 454 (1969).

[17] A. Trisorio, S. Grabielle, M. Divall, N. Forget, and C. P. Hauri, *Self-referenced spectral interferometry for ultrashort infrared pulse characterization,* Opt.Lett. 37, 2892{2894 (2012).

[18] Perego, C. et al., 2011. *Extensive comparison among Target Normal Sheath Acceleration theoretical models.* Nuclear Instruments and Methods in Physics Research, Section A: Accelerators, Spectrometers, Detectors and Associated Equipment, 653(1), pp.89–93.

[19] Neely, D. et al., 2016. Appl. Phys. Lett. 89, 021502 (2006); doi: 10.1063/1.2220011.

[20] Maksimchuk, a et al., 2000. *Forward ion acceleration in thin films driven by a high-intensity laser.* Physical review letters, 84(18), pp.4108–11. Available at: http://www.ncbi.nlm.nih.gov/pubmed/10990622.

[21] S. Fourmaux, S. Buffechoux, B. Albertazzi, D. Capelli, A. Levy, S. Gnedyuk, L. Lecherbourg, P. Lassonde, S. Payeur, P. Antici, H. Pepin, R. S. Marjoribanks, J. Fuchs, and J. C. Kieffer, Phys. Plasmas 20, 013110 (2013).

[22] Schnürer, M., 2011. *Comparison of femtosecond laser-driven proton acceleration using nanometer and micrometer thick target foils.* Laser and Particle Beams, 29(4), pp.437–446. Available at: http://dx.doi.org/10.1017/S0263034611000553.

[23] P. McKenna, F. Lindau, O. Lundh, D. Neely, A. Persson, and C.-G. Wahlstrom, *High-intensity laser-driven proton acceleration: Inuence of pulse contrast,*Philosophical Transactions: Mathematical, Physical and Engineering Sciences 364, 711{723 (2006).

[24] Kiefer, T., 2014. Investigation of the laser-based Target Normal Sheath Acceleration (TNSA) process for high-energy ions an analytical and numerical study.

IOP Conf. Series: Journal of Physics: Conf. Series **1079** (2018) 012001 doi:10.1088/1742-6596/1079/1/012001

[25] Fuchs, J. et al., 2006. Laser-driven proton scaling laws and new paths towards energy increase.Nature Physics, 2(1).

[26] Zaffino R. L, Sanchez I, Ruiz A, Seimetz M, Quirion D, Lera R, Bellido P, Benlloch JM, Lozano M, (2017) Wafer scale fabrication of high density targets array for stable generations of protons beams by laser-plasma interaction, J Phys Conf S (to be published).

[27] Zaffino R.L., Seimetz M., Quirión D, Ruiz A., Sánchez I., Mur P,, Benlliure J., Martín L., Roso L., María J., Benlloch P, Lozano M, Pellegrini G, *Preparation and characterization of micro-nano engineered targets for high-power laser experiments*, Microelectronic Engineering, Vol 194, 2018, p.p 67-70, ISSN 0167-9317.

[28] Seimetz M, Bellido P, Soriano Asensi A, Conde Castellanos PE, Crespo Navarro E, González Martínez AJ, Hernández Hernández L.(2015). *Calibration and Performance Tests of Detectors for Laser-Accelerated Protons.* IEEE Transactions on Nuclear Science. 62(6):3216-3224.

Acknowledgements

The authors wish to thank the technical staff of the UPC2 Instituto de Microelectrónica de Barcelona IMB-CNM (CSIC), Spain, for their technical assistance in fabrication of the test samples, and to the Instituto de Instrumentación para Imagen Molecular (I3M) for the TOF design and calibration. This work has been partially supported by the Spanish Ministry for Economy and Competitiveness within the Retos-Colaboración, ref. RTC-2015-3278-1.

IOP Conf. Series: Journal of Physics: Conf. Series **1079** (2018) 012002 doi:10.1088/1742-6596/1079/1/012002

Integrated double-plasma-mirror targets for contrast enhancement in laser ion acceleration

Martin Speicher[1], Daniel Haffa[1], Matthias A. O. Haug[1], Jianhui Bin[1,2], Ying Gao[1], Jens Hartmann[1], Peter Hilz[1], Christian Kreuzer[1], Florian H. Lindner[1], Tobias M. Ostermayr[1,3], Thomas F. Rösch[1], Rong Yang[1], Jörg Schreiber[1,3]

[1]Department of Medical Physics, Faculty of Physics, Ludwig-Maximilians-Universität München, Am Coulombwall 1, 85748 Garching b. München, Germany
[2]Accelerator Technology and Applied Physics Division, Lawrence Berkeley National Laboratory, Berkeley, California 94720, USA
[3]Max-Planck-Institut für Quantenoptik, Hans-Kopfermann-Straße 1, 85748 Garching b. München, Germany

Email: martin.speicher@physik.uni-muenchen.de

Abstract. We introduce a target concept for laser-driven ion acceleration with ultrashort, highly-intense laser pulses that includes an integrated double plasma mirror for contrast enhancement. It comprises three nanometer thin plastic foils, embedded in a small metal structure, which ensures precise mounting. The geometry allows to apply a double plasma mirror directly in front of a target foil within the converging beam, enabling moderate-energy (~1 J) laser systems to reach the required fluence of several hundred J/cm^2 on the plasma mirrors. During an experimental campaign, performed at the Laboratory for Extreme Photonics in Munich, we observed proton energies increased by a factor of three using this new target, which is attributed to an enhanced laser contrast after the integrated double plasma mirrors.

1. Introduction

In the upcoming Petawatt era, laser facilities all over the world are targeting to reach peak intensities up to 10^{22} W/cm^2 in a focal spot of a few micrometers diameter [1]. At such extreme conditions, the target's ionization threshold (~10^{13} W/cm^2) is exceeded by many orders of magnitude and precise knowledge and control of the temporal laser profile (laser contrast) is often crucial for experimental performance. Pre-pulses, that originate from the amplification process of the laser, precede the main pulse on ns- to fs-timescales. This can cause a pre-expansion of the target and can alter the often desired sharp vacuum-solid boundary, therefore strongly influencing experimental results [2–4].

Several contrast-cleaning techniques have been developed over the last years to suppress such pre-pulses. The ASE-level can be lowered by optical parametric chirped pulse amplification (OPCPA) [5,6], saturable absorbers [7] or cross-polarized wave generation (XPW) [8], whereas single pre-pulses can be removed to some extent by fast Pockels cells and the avoidance of transmissive optics in the laser. All these components can improve the laser contrast but also partially require major reconstructions in the laser setup.

A self-triggering plasma mirror (PM), which usually consists of a fused silica substrate with an anti-reflection coating [9–13], provides a simple means of contrast improvement by orders of magnitude and is usually implemented after the compression of the amplified laser. To adjust its ignition, the PM is placed along the converging laser, usually before focus, such that the intensity of any pre-pulse remains below the ionization threshold of the substrate, whereas the rising part of the main pulse generates a reflective plasma. With a reflectivity of ~10^{-2}-10^{-3} for the pre-pulse and 0.6 – 0.8 for the main pulse, a single

IOP Conf. Series: Journal of Physics: Conf. Series **1079** (2018) 012002 doi:10.1088/1742-6596/1079/1/012002

PM enables significant contrast enhancement of two to three orders of magnitude, at the expense of laser energy on target.

A larger number of plasma mirrors can be employed to further improve contrast. The use of a double plasma mirror (DPM) has in fact several motivations. In a re-collimating setup, the ideal fluence for triggering the plasma mirror is available before and after focus, offering twice the contrast enhancement in a single telescope setup. The angle of incidence, however, is around 45° then and this has detrimental effects on the main pulse reflectivity [14]. Nevertheless, the use of such DPMs *after* the final focusing parabola has proven very useful, because it avoids reflecting the laser over yet more optics that potentially contribute to contrast deterioration and, more pragmatically, does not require re-arranging the complete experiment.

Proper reflection requires a certain fluence (several 10s to 100s J/cm^2) [15], hence in a DPM used in a converging beam the main pulse reflectivity can never be as optimal as in two independently operated single PMs (or a telescope setup). So far, the use of DPMs as described had been limited to large scale systems with kJ energy and large beam diameters [16]. Such pulses can trigger the PM centimeters before focus, thus providing sufficient space for their implementation prior to target. For laser systems in the low-J range, the distance to target reduces to some millimeters, which has so far hindered the use of this generally simple concept at such systems.

We present a new compact target design, consisting of three nanometer-thin plastic foils, embedded in a metal structure, where two of these foils function as a DPM and the third as the target for laser-ion acceleration. Using this approach, the DPM can be placed as close as 2.9 mm and 0.9 mm to the target. Therefore, the new design provides a small, cost-efficient setup for significant contrast enhancement, which allows implementation in particularly tight experimental setups.

2. Concept and production

Our DPM target was designed for an f/2 off-axis parabolic mirror and 2 J laser energy on target with horizontal polarization (p-polarized in the PM-case). A sketch of the DPM geometry and the incident laser is shown in figure 1a, where $\alpha = 14°$ is the convergence angle of the focused beam. The design was optimized for achieving the required fluence on the first mirror, minimizing the path to the second and preventing clipping at the metal structure that supports the thin foils (parameters x and m).

Feasible values for all geometric parameters were determined by numerical analysis and yielded an optimum incidence angle of $\beta = 30°$, resulting in fluences of 100 J/cm^2 on the first and 800 J/cm^2 on the second PM. Those fluences lead to intensities of $4 \cdot 10^{15}$ W/cm^2 and $3.2 \cdot 10^{16}$ W/cm^2 on PM1 and PM2, respectively. We therefore estimate a total reflection of $R_{calc,high} = 0.675 \cdot 0.68 = 0.46$ for a laser pulse with 25 fs duration [9].

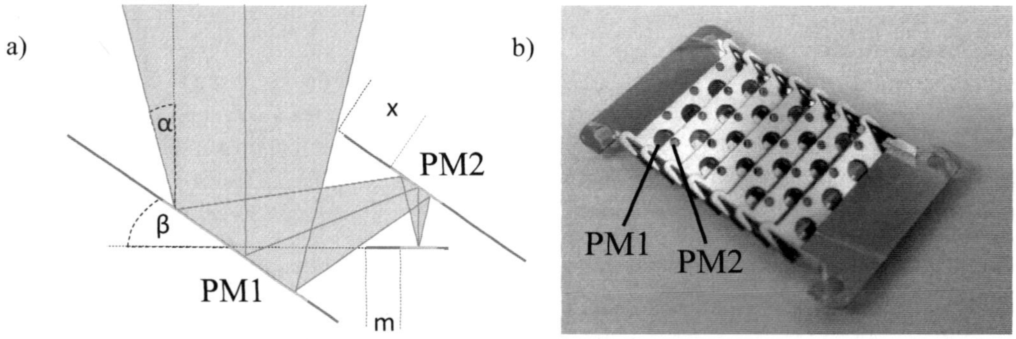

Figure 1. a) This sketch shows the geometry of the DPM target. The path of the laser is illustrated in blue. The design parameters are the opening angle β and distances x and m in order to prevent clipping at the metal structure. **b)** Image of the final target holder containing 18 DPM targets.

Optimization of the PM performance (i.e. contrast enhancement) can be achieved by minimizing the reflectivity for pre-pulses. This is done by matching the PM foil thickness to the destructive interference criterion for the central laser wavelength of $\lambda = 800$ nm: $d_{pm} = \lambda / (2 \cdot \cos \delta \cdot n_{pm}) = 295$ nm, where $\delta = 20°$ is the light angle in the PM foil and $n_{pm} = 1.445$ the refractive index of the PM material. The foils used for the PM had a thickness of (270 ± 30) nm. The reflectivity for a single PM can be calculated using the formula for thin film etalons: $R = 1 - 1 / (1 + F \sin^2(2 \cdot \pi \cdot n_{pm} \cdot d \cdot \cos(\delta) / \lambda))$, where $F = 4 \cdot R_p / (1-R_p)^2$ and R_p is the reflectivity on the foil frontside for p-polarized light. For a thickness d = 270 nm, one obtains R = 0.0057. But the reflectivity is also influenced by the fact that the laser is strongly convergent ($\Delta\beta = 14°$) and broad band ($\Delta\lambda = 50$ nm). Therefore, $R_{min} = [((\partial R/\partial\lambda) \cdot \Delta\lambda)^2 + ((\partial R/\partial\beta) \cdot \Delta\beta)^2]^{1/2} = 0.011$ will represent a lower limit in reflectivity of a single pass. The DPM reflectivity therefore is $R_{calc,low} = (R + R_{min})^2 = 0.279 \cdot 10^{-3}$ and we hence estimate a contrast enhancement of $R_{calc,high} / R_{calc,low} = 1648$.

The PM foils are produced in a quick and cost-efficient way using the droplet method [17], for which Formvar serves as the PM material due to its high reproducibility. The third foil, in our study functioning as the target for laser ion acceleration, can be chosen as demanded by experimental requirements, for example regarding its thickness or material. The foils are applied on several metal plates, which are then patched to achieve the geometry described above. The angular mounting accuracy minimizes the deviation of beam pointing (final focus) within a square of 200 µm × 200 µm, i.e. within the field of view of the focus diagnostic microscope. As a final result, a metal frame offers a compact setup with a total number of 18 DPM targets (see figure 1b).

3. Experimental implementation and results

The DPM system was implemented at the ATLAS 300 laser at the Laboratory for Extreme (LEX) Photonics at LMU Munich in Garching. An attenuated beam was used to determine the reflectivity of the DPM setup without target foil. Comparing the intensity of transmitted light with and without PM yielded $R_{meas,low} = 0.0032 \pm 0.001$, which is 10 times larger than estimated above. It is most likely that this high reflectivity is a result of insufficiently well controlled PM thickness, for example a foil with d = 240 nm would theoretically have a similar reflectivity. The same procedure with full power resulted in $R_{meas,high} = 0.49$, which is in good agreement with the theoretical value. The measured contrast enhancement for our DPM system is therefore $R_{meas,high} / R_{meas,low} = 153$.

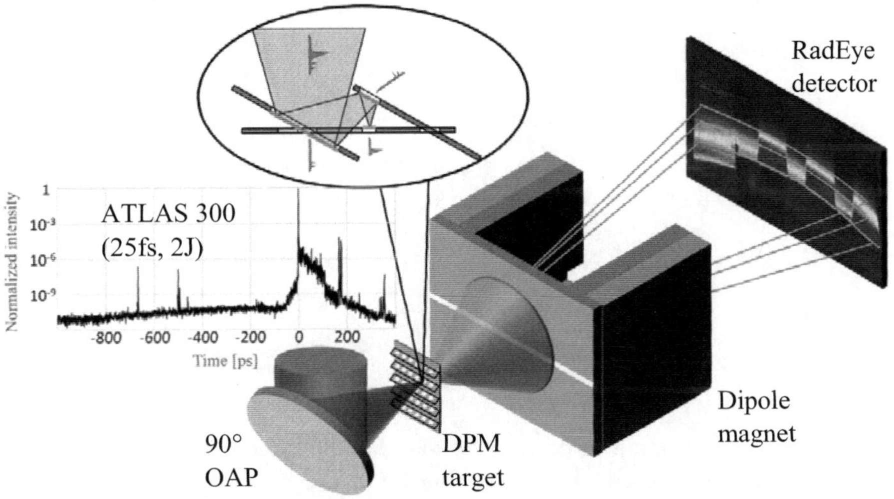

Figure 2. Schematic of the experimental setup: a 90° off-axis parabola focuses the laser onto the DPM target. After passing a dipole magnet, the accelerated protons are recorded by a RadEye detector. The functionality of the DPM and an exemplary laser contrast curve of the incident laser is pictured in the inset.

IOP Conf. Series: Journal of Physics: Conf. Series **1079** (2018) 012002 doi:10.1088/1742-6596/1079/1/012002

The schematic of the experiment used for laser ion acceleration with the described DPM target is shown in figure 2. The incident laser is focused onto the target by an f/2, 90° off-axis parabola. The pulses contain 2 J of energy within a pulse duration of 25 fs. An exemplary contrast curve shows two short pre-pulses at 670 and 500 ps prior to the main pulse with a contrast ratio of almost 10^{-6}. Considering a peak intensity of 10^{20} W/cm^2, both pre-pulses exceed the damage threshold of the target material. Although such pre-pulses can be suppressed in principle in the laser, this situation serves well for our demonstration study. The DPM reduces the pre-pulses by more than a factor of 100, sufficient to suppress their intensity to below the damage threshold of typical target materials. The target frame was constructed compatible to and mounted onto the inhouse-built nano-foil target positioning system [18] (not shown in figure 2). The emitted protons are deflected by a dipole magnet and recorded by online RadEye detectors. The detector is covered with different layers of aluminum (45 μm on the left side, 75 μm on the right) in order to define energies at 2.1 and 2.9 MeV for subsequent calibration and evaluation of the detected protons energy distribution [19].

Figure 3 shows experimental results, with quantitative differential spectral amplitudes of the protons plotted against the kinetic energy. The red curve represents a typical shot on a single plastic foil of 500 nm thickness (without PM, i.e. no contrast enhancement), whereas blue, yellow and green mark shots on DPM-targets with 20 nm thick plastic foils. The noise level of each shot is shown by the dashed lines in the corresponding color. The maximum proton energies are defined as the intersection point of signal and noise for every shot. Looking at these representative shots we can already see an increase of the kinetic energy up to 11 MeV and differential spectral amplitudes by up to two orders of magnitude if we compare the DPM target with a single plastic foil. For the following we extract only the maximum proton energy for the evaluation of the acceleration performance with and without plasma mirrors.

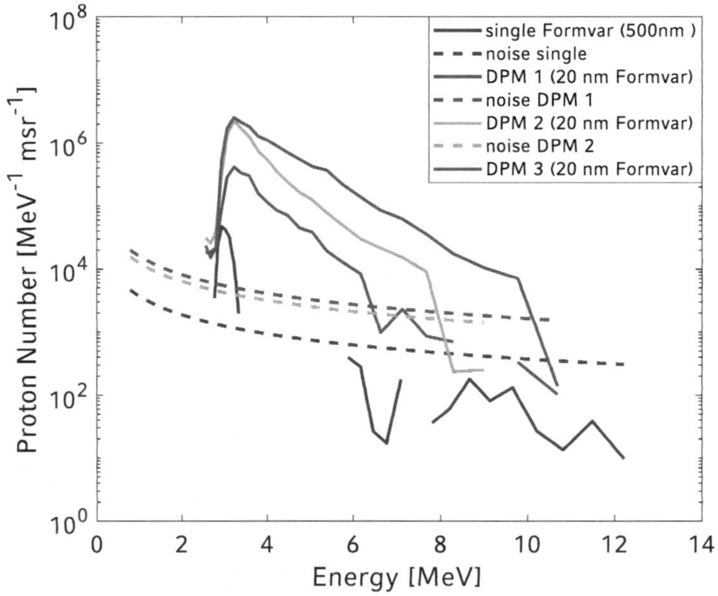

Figure 3. Exemplary proton spectra, i.e. differential spectral amplitudes of the protons are plotted against kinetic energy for four exemplary shots.

Figure 4 summarizes maximum proton energies for different target materials and thicknesses. Circles mark energies which were acquired using the DPM target, whereas triangles symbolize shots without contrast enhancement. Shots with plastic targets (Formvar and polystyrene) were evaluated for two experimental days in order to provide information on reproducibility of the results. On one of the days, diamond-like carbon (DLC) was used as target material as well. The DPM foils were exclusively 270 nm thick Formvar for all shots.

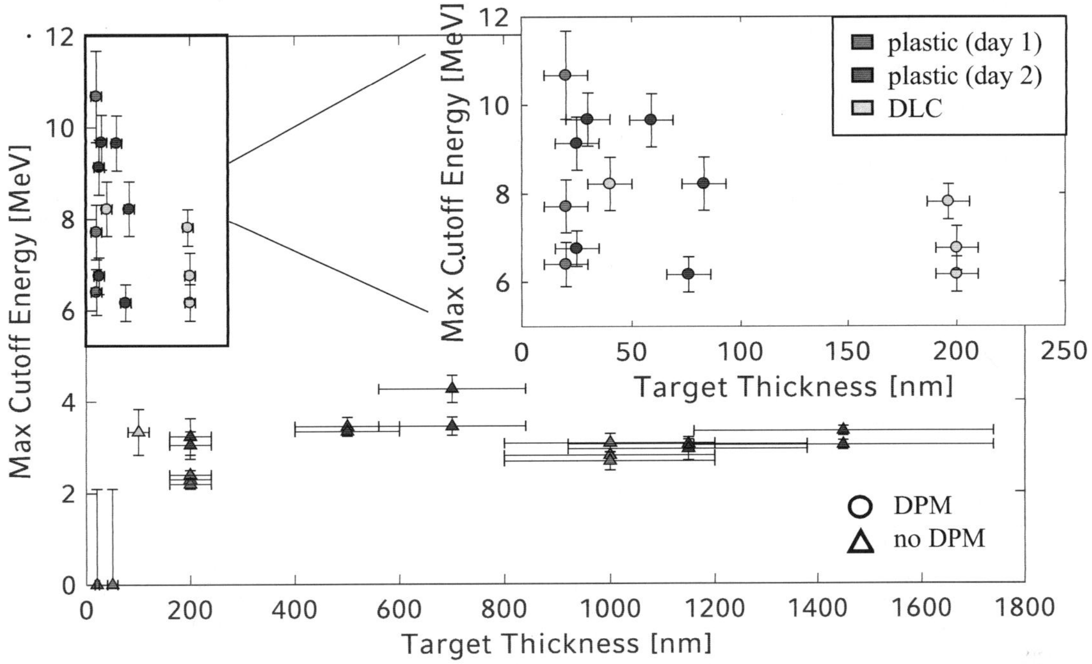

Figure 4. The plot shows the maximum proton energies for different target materials and thicknesses. Circles mark shots with DPM targets, whereas triangles symbolize single foil targets without DPM. Vertical error-bars resemble the detector resolution, horizontal error-bars cover a thickness fluctuation of ± 10 nm for DPM targets and ± 20 % for single foils.

When using single plastic foils, the minimum target thickness for detectable ions above 2.1 MeV appeared to be 200 nm, for DLC it was 100 nm. The highest proton energy of 4.2 MeV was observed for 700 nm thick targets. With 2 J on target, those numbers are clearly insufficient. Exemplary snapshots of the laser light transmitted through 20 nm and 200 nm thin targets shot without PM (Fig 5a and b) suggest that the collapse of proton acceleration from 20 nm foils is accompanied by target transparency. This is likely due to pre-expansion induced by short pre-pulses that arrive long before the main pulse [20]. Although the pulse energy is reduced to ~1 J when inserting the DPM, proton energies increase to values between 6 - 11 MeV at foils of 20 nm thickness. In this case, the target remains largely opaque (Fig 5c).

Figure 5. Exemplary images of a transmission screen behind the target, where **a)** shows a shot on a single 20 nm thick plastic foil without PM (96.8% transmission), **b)** a shot on a single 200 nm thick plastic foil without PM (11.3% transmission) and **c)** a shot with DPM and a 20 nm target foil (14.9% transmission)

IOP Conf. Series: Journal of Physics: Conf. Series **1079** (2018) 012002 doi:10.1088/1742-6596/1079/1/012002

4. Summary and discussion

We have demonstrated a new target concept that enables accommodating a DPM for laser ion acceleration close to the focus. This integrative design is particularly beneficial for laser systems with energies in the low-J regime. The demonstrated solution enables precise positioning via implementation in existing target systems [18]. Foils can be fabricated cost-efficient with high reproducibility in advance to experimental campaigns. A contrast enhancement of more than two orders of magnitude is viable, in line with successful proton acceleration from targets as thin as 20 nm. Future improvement must concentrate on better control of the PM thickness and homogeneity as well as an on-shot/single shot diagnostic for intensity distribution measurements to reduce the currently observed large shot-to-shot variation. Although based on non-optimal contrast, our concept may provide a quick solution to increase the proton energy by more than a factor of two and the proton numbers by more than a factor of 100 without having to intervene into the laser system, despite the reduced energy on target of 0.98 J.

Remaining challenges comprise mainly the improvement of mounting accuracy. We are currently operating the system in shot-on-demand mode as the limited angular accuracy of the mounting leads to fluctuations of the focus position inside a 200 µm × 200 µm square which requires realignment for every shot. Although this fact limits our repetition rate up to now, target mass production prior to experiment is already possible which would favor a high-repetition system if the positioning accuracy is improved. Due to the fast production and good reproducibility we choose the droplet method for the foil fabrication. Recent measurements have shown though that a better foil quality can be achieved by using spin-coated foils which would not only improve the beam quality but also reduce the reflectance of the low intensity part as thickness fluctuations would decrease. As a last point it has to be mentioned that, in the current setup, foils close to the primary focus are frequently destroyed. In the worst case, only 4-8 of the 18 targets could be used in our experiment. Mitigating this effect will be essential to benefit from the full capacity in future studies.

Acknowledgement
The DFG-funded Cluster of Excellence Munich-Centre for Advanced Photonics (MAP) supported this work. We thank the support of the operation team of the ATLAS laser system. F.H.L. was supported by the BMBF under contract 05P15WMEN9. T.F.R. is scholar of the German Academic Scholarship Foundation. J. H. B. is funded by a Feodor Lynen Fellowship of the Alexander von Humboldt Foundation.

Author contributions
D. H., C. K., F. H. L., T. O. and P. H. built the experimental infrastructure in LEX-Photonics and designed the experiment setup. M. A. O. H., D. H., P. H. and J. S. designed and developed the DPM system. M. S., D. H., J. H., J. H. B., T. O., Y. G., R. Y., F. H. L., T. F. R. and C. K. performed the experiment at LEX-Photonics. M. S., J. H. B., D. H. and J. S. analyzed the data, discussed and interpreted the results and prepared the manuscript.

References
[1] Danson C, Hillier D, Hopps N, Neely D 2015 *High Power Laser Sci. Eng.* **3**
[2] Kaluza M, Schreiber J, Santala M I K, Tsakiris G D, Eidmann K, Meyer-ter-Vehn J, Witte K J 2004 *Phys. Rev. Lett.* **93**(4) 045003
[3] McKenna P, Lindau F, Lundh O, Neely D, Persson A, Wahlstrom C G, 2006 *Phil. Trans. R. Soc. A* **364** 711–23
[4] Bagnoud V, Hornung J, Schlegel T, Zielbauer B, Brabetz C, Roth M, et al 2017 *Phys. Rev. Lett.* **118**(25)

IOP Conf. Series: Journal of Physics: Conf. Series **1079** (2018) 012002 doi:10.1088/1742-6596/1079/1/012002

[5] Dubietis A, Jonušauskas G, Piskarskas 1992 *Opt. Commun.* **88**(4–6) 437–40
[6] Ross I, Matousek P, Towrie M, Langley A J, Collier J L 1997 Opt. Commun. **144**(1–3) 125-33
[7] Fourmax S, Payeur S, Buffechoux S, Lassonde P, St-Pierre C, Martin F, Kieffer J C 2011 *Optics Express* **19**(9) 8486-97
[8] Jullien A, Albert O, Burgy F, Hamoniaux G, Rousseau J-P, Chambaret J-P, et al 2005 *Opt. Lett.* **30**(8) 920-2
[9] Poole P L, Krygier A, Cochran GE, Foster P S, Scott G G, Wilson L A, et al 2016 *Sci. Rep.* **6** 32041
[10] Thaury C, Quéré F, Geindre J-P, Levy A, Ceccotti T, Monot P, et al 2007 *Nat. Phys.* **3**(6) 424-9
[11] Kapteyn H C, Szoke A, Falcone R W, Murnane M M 1991 *Opt. Lett.* **16**(7) 490-2
[12] Backus S, Gold D M, Nathel H, Kapteyn H C, Murnane M M, White W 1993 *Opt. Lett.* **18**(2) 134-6
[13] Dromey B, Kar S, Zepf M, Foster P 2004 *Rev. Sci. Instrum.* **75**(3) 645–9
[14] Ziener C, Foster P S, Divall EJ, Hooker C J, Hutchinson M H R, Langley AJ, et al 2003 *J. Appl. Phys.* **93**(1) 768–70
[15] Doumy G, Quéré F, Gobert O, Perdrix M, Martin P, Audebert P, et al 2004 *Phys. Rev.* E **69**(2)
[16] Dromey B, Zepf M, Gopal A, Lancaster K, Wei M S, Krushelnick K, et al 2006 *Nat. Phys.* **2**(7) 456–9
[17] Seuferling S, Haug M A O, Hilz P, Haffa D, Kreuzer C, Schreiber J 2017 *High Power Laser Sci. Eng.* **5**
[18] Gao Y, Bin J H, Haffa D, Kreuzer C, Hartmann J, Speicher M, et al 2017 *High Power Laser Sci. Eng.* **5**
[19] Lindner F H, Bin J H, Englbrecht F, Haffa D, Bolton P R, Gao Y, et al 2018 *Rev. Sci. Instrum.* **89**, 013301
[20] Zhou ML, Bin J H, Haffa D, Yan XQ, Schreiber J 2017 *Plasma Phys. Control. Fusion* **59**(5) 055020

IOP Conf. Series: Journal of Physics: Conf. Series **1079** (2018) 012003 doi:10.1088/1742-6596/1079/1/012003

The spatial contrast challenge for intense laser-plasma experiments

Jens Hartmann[1], Daniel Haffa[1], Martin Speicher[1], Jianhui Bin[2], Peter Hilz[1], Christian Kreuzer[1], Tobias Ostermayr[1,3], Sebastian Lehrack[1], Jörg Schreiber[1]

[1]Lehrstuhl für Medizinphysik, Fakultät für Physik, Ludwig-Maximilians-Universität München, 85748 Garching b. München, Germany
[2]Accelerator Technology and Applied Physics Division, Lawrence Berkeley National Laboratory, Berkeley,
CA 94720, USA
[3]Max-Planck-Institut für Quantenoptik, 85748 Garching b. München, Germany

Email: jens.hartmann@physik.uni-muenchen.de

Abstract. Achieving the highest peak intensity possible is a key requirement of a majority of laser-plasma experiments. This requires confining laser energy to a small volume. The experimental determination of the spatial focus distribution is therefore important, especially to connect theoretical and experimental results. In this paper, we revise a new method to evaluate the spatial performance of a laser focus. Obtaining low- and high-dynamic-range (LDR/HDR) images of the laser focus profile is described for a typical high-intensity laser setup, and present data from our case study. We compare standard evaluation procedures based on both images quantitatively and determine the accuracies and associated errors. A spread out intensity distribution below the detection limit of the LDR is observed in the HDR acquisition. Our evaluation reveals that only considering LDR can overestimate the enclosed energy as well as the peak intensity by factors of 1.5. With our method we estimate the required dynamic range for this case to 4 orders of magnitude. Our discussion sheds light on possible causes for the deteriorative effect to targets in the vicinity of the primary focal spot.

1. Introduction

Laser systems exceeding one peta-watt (PW) peak power have started operation or are currently under construction. Achieving intensities beyond $10^{21}\,\mathrm{W}cm^{-2}$ will become possible on a regular basis, which is of interest for applications of laser-plasma interactions, such as laser-driven ion acceleration [1] , laser-driven electron acceleration [2] or high harmonic generation [3]. An accurate measurement of the laser fluence profile in the target plane is typically used to derive a spatial intensity distribution, peak intensity and the energy enclosed in focus. These quantities are essential for simulations and analytical models that supplement experimental data and steer future experiments. These laser-parameters have also particular impact on establishing and extrapolating analytical models and scaling.

Peak intensity of a laser focus might be in future determined by measuring the phase after the focus [4] or by measuring the ionization induced by the laser field [5]. The most common method is based on imaging the focal plane onto a camera chip [6] and recording the attenuated, time-integrated laser fluence distribution $F(x, y)$, which is typically normalized $\int F(x, y)dxdy = 1$. Analogously, the temporal intensity distribution $f(t)$ is usually obtained from Frequency Resolved Optical Gating (FROG) measurements [7], providing typically a few orders of magnitude dynamic range with fs resolution, or autocorrelation measurements [8], offering many orders of magnitude dynamic range with typically ~>100 fs resolution. Assuming the spatio-temporal intensity distribution to be separable and I_0 as the peak intensity, $I(x, y, t) = I_0 F(x, y) f(t)$. A Gaussian or a sinc² distribution can be fitted to the data, and the full width at half maximum distributions (FWHM) d_x, d_y and τ obtained.

In practice it is not necessary to make an assumption for the distribution function. Instead, the focal spot image is numerically evaluated and required to fulfill $\int I_0 F(x, y) f(t) dxdydt = E_L$, with E_L being the laser energy. It is immediately obvious that the dynamic range with which the spatial (and temporal) distribution is obtained will influence the determination of the peak intensity. Here, we concentrate on evaluating the spatial intensity distribution but note, that the methodology can be transferred to the temporal domain for low temporal contrast systems.

2. Case Study

The data have been acquired at the Advanced Ti:Sa LASer system (ATLAS), during operation at the Laboratory of Extreme Photonics (LEX-Photonics) in Garching bei Muenchen, Germany. The system runs at a central wavelength of 800 nm and is capable of delivering 25 fs pulses with 2 J on target at a repetition rate of up to 5 Hz.

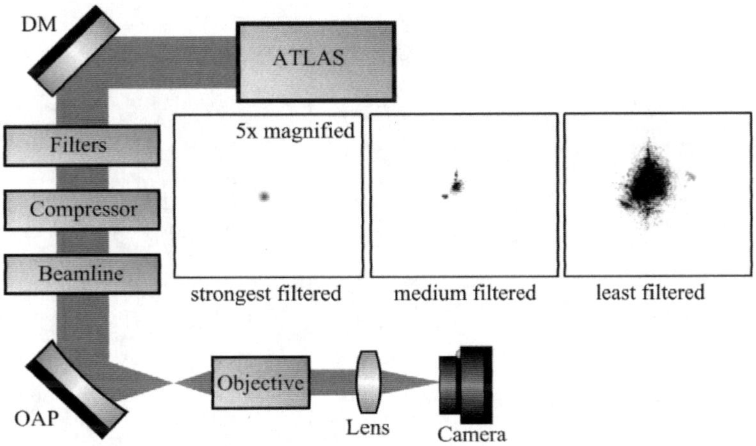

Figure 1. Case study setup. Raw images acquired with different filter settings are shown in the inset. The strongest filtered image is magnified 5 times relative to the others. Figure made using [9].

Figure 1 shows the experimental setup. Prior to re-compression, the wave-front of the pulse is corrected by a deformable mirror (DM). The laser beam can be attenuated with a combination of neutral-density filters with extinction ratios of 0.3, 0.5 and 1 order of magnitude, and a dielectrically coated mirror filter with an extinction of 7 orders of magnitude. It is then guided via six mirrors through an evacuated laser beam delivery into the experimental chamber, in which a f/2, $f = 20cm$, 90° off-axis parabolic mirror (OAP) focusses the laser. A vacuum compatible magnification images the focus onto the complementary metal–oxide–semiconductor (CMOS) camera chip UI-5241-LE [10], with an analog-digital conversion of 10-bit corresponding to a dynamic range of three orders of magnitude, assuming ideal signal to noise ratios.

IOP Conf. Series: Journal of Physics: Conf. Series **1079** (2018) 012003 doi:10.1088/1742-6596/1079/1/012003

Focal spot images are recorded with an attenuated laser beam. The filters are set such, that the dynamic range provided by the camera is well exploited. We define such an image, if it has less than three orders of dynamic range, as low-dynamic-range (LDR).

The high-dynamic-range (HDR) approach starts with this LDR-setting. The filters are then removed stepwise and a picture is captured for each attenuation. This leads to a deliberate saturation of the focus center, but lifts outer parts beyond the signal-to-noise-ratio (SNR). The images are stacked on top of each other, where slight lateral shifts due to focus jitter or filter usage have to be accounted for. This is done manually for each picture. The contours of line-out distributions for each spatial dimension (such as Figure 2 c) through the image center are compared, and shifted until the best match is achieved. With respect to the previous image, I5 has been shifted 0.8 µm, I6 2.4 µm and I7 17 µm.

As a result of the stacking, the pixels which have been overexposed in the lesser filtered image are replaced by the accordingly scaled values in the follow up (stronger filtered) image. The dynamic range is hence extended in our case by two orders of magnitude. It is practically limited by laser-induced damage (or blooming) of the camera chip.

A line-out of the HDR fluence distribution is shown in Figure 2 c) to demonstrate the process of stacking the images. I1 contains only the information gathered by the lowest filter setting, I2 contains I1 and a second image and so forth. I7 is the final result and is referred to as HDR focus. The dashed blue line indicates the detection limit of the camera chip. It is scaled with the same factors used for stacking the image data on top of each other. The two-dimensional HDR distribution is shown in Figure 2 a).

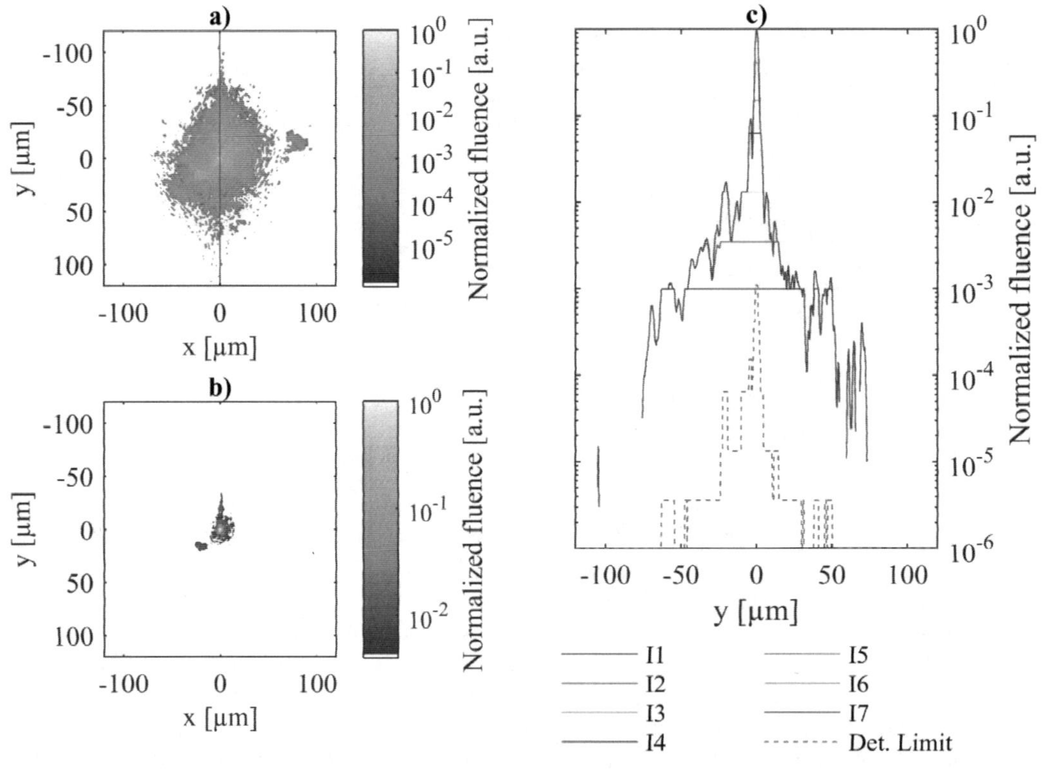

Figure 2. a) Reconstructed HDR fluence distribution of the laser focal spot, normalized to a maximum of 1. b) Apparent LDR image as it would be observed by a 8-bit camera. c) Line-out through the maximum along the vertical red axis of the HDR-image in a)

IOP Conf. Series: Journal of Physics: Conf. Series **1079** (2018) 012003 doi:10.1088/1742-6596/1079/1/012003

In order to suppress noise, a constant pixel value of 47 has been subtracted from the picture, where pixel values < 0 have been set to zero. This results in 95% of the maximum dynamic range of this camera chip being used, which is nearly approaching three orders of magnitude. The indicated detection limit in Figure 2 c) is the equivalent of one pixel value, multiplied with the same scaling factor applied to the corresponding pixel in the fully stacked imaged I7.

In order to compare the differences between the HDR and LDR method, a LDR picture (Figure 2 b) has been derived from the HDR image (Figure 2 a), by setting pixels with values $P(x, y) < 10^{2.4}$ to zero, thus simulating a typical 8-bit camera acquisition.

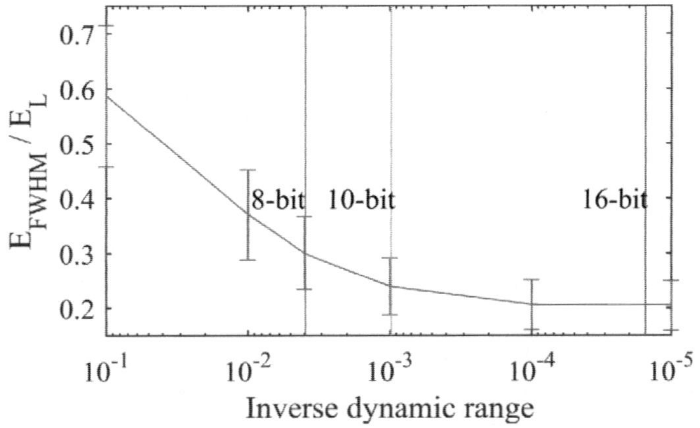

Figure 3. Energy enclosed in FWHM for evaluations considering a hypothetical dynamic range of the employed camera-chip for focal spot imaging. Dynamic ranges corresponding to typical linear camera A/D converter bit depths are indicated. The error bars indicate values at +/- 5% of the half maximum.

Direct comparison of the HDR and LDR fluence distributions in Figure 2 a) and b) reveal the important additional information obtained by the HDR approach. The lower detection threshold indicates a pedestal with ~ 100 µm in diameter. In order to quantify the impact of detecting this pedestal on determining essential laser pulse parameters, we devised the following method: Using the same procedure that yielded Figure 2 b), we generated images at dynamic ranges from 10^1 to 10^5. The energy enclosed in the FWHM diameter $E_{FWHM} = E_{>0.5} \,/\, E_L = \int F_{>0.5}(x, y)dxdy$ has been evaluated for each range and the result plotted in Figure 3. The graph is converging at an inverse dynamic range of 10^{-4}, therefore a single shot image by a camera with 16-bit should suffice to extrapolate meaningful laser parameters for this particular case.

The peak intensity was calculated by dividing the largest fluence value in the processed focus image by the FWHM pulse duration, with $I_0 = 3.3 * 10^{20} W/cm^{-2}$ for the HDR case, and $4.8 * 10^{20} W cm^{-2}$ in the LDR case. Considering the LDR only would hence have introduced a systematic overestimation of peak intensity by a factor of 1.5.

3. Discussion

We described a procedure to acquire HDR and LDR laser focus fluence images. Evaluating the HDR image allows to simulate the effect of insufficient dynamic range. In our specific case, a dynamic range of 16-bit seems sufficient for capturing the majority of the energy. We note though, that this procedure is not mathematically strict. It would still allow for a very wide spread contribution at even lower intensity that we would miss. Ideally one should perform an additional measurement to obtain the energy fraction that is indeed contained within the area were the signal is larger than the noise, for example by introducing an appropriate aperture.

We recorded the spatial intensity distribution over five orders of magnitude. Scaled to the estimated peak intensity, the current detection limit still remains as high as $10^{15} W cm^{-2} \gg 10^{13}\ W cm^{-2}$, which is the damage-threshold of solid-density plastic targets used in laser-plasma experiments. Our pedestal already has a diameter of ~ 140 μm, and it is uncertain if an even larger pedestal below our HDR detection threshold exists.

It is worth noting that, during laser-driven ion acceleration experiments with solid-density plastic targets, we compared shots with target foils facing the laser, with shots where the foil was on the non-irradiated side of the target-holder. For the laser-side case, we observed mm-scale burn spots, whereas if the foil was shielded by the target-holder, the damage to the foil is contained to a small radius around the target-hole. It will be interesting to determine the laser focus pedestal to even lower intensities in the future, in order to investigate potential correlation with target damage.

Assuming a similar focus performance, the transition to even higher intensities leads to significantly larger areas around the laser focus to be exposed to laser induced damages. Depending upon design, solid-density or gas-nozzle target systems might suffer from collateral laser-damage. In the near future achieving laser peak-intensities by focus optimization up to $10^{24} \mathrm{W} cm^{-2}$ is feasible. As this is eleven orders of magnitude larger than the damage threshold of current targets, monitoring this spatial contrast requires devices of very high dynamic range and therefore is going to become a challenge. Camera chips capable of more than four orders of dynamic range have been developed [11] and are commercially available.

Achieving high intensity requires both good spatial and temporal confinement of the laser energy. The influence of insufficient temporal contrast, for example due to amplified spontaneous emission and coherent contrast is critical to most state-of-the-art experiments, but not necessarily through is impact on estimating the peak intensity. Due to the similarities with our observation, we refer to the topic of this work as the spatial contrast challenge.

For efficient use of PW-lasers, it will be important to identify the sources of the degradation of this spatial contrast. Amongst possible candidates are imperfections in laser-beam transport or final focusing optics. Irregularities on a sub-mm scale are currently not very well identifiable by established means such as interferometry. Using the described method to qualify optics with large aperture, in particular ones with strong curvature (low f# off-axis parabolic mirrors) could become an important tool complementing interferometric surface analysis.

Author Contribution

D. H., C. K., T. O., P. H. and S. L. built the experimental infrastructure in LEX-Photonics and designed the experiment setup. D. H., J. H., M. S., J. B., performed the experiment at LEX-Photonics. J.H., J. S.,M.S., and J. B. analyzed the data, discussed and interpreted the results and prepared the manuscript.

Acknowledgements

The DFG-funded Cluster of Excellence Munich-Centre for Advanced Photonics (MAP) supported this work. We thank the ATLAS laser operation team around Stefan Karsch.

References

[1] Macchi A, Borghesi M and Passoni M 2013 *Rev. Mod. Phys.* **85**, 751–793
[2] Lefebvre E et al. 2003 *Nucl. Fusion* **43**, 629
[3] Teubner U and Gibbon P 2009 *Rev. Mod. Phys.* **81**, 445–479
[4] Bahk S-W et al. 2004 *Opt. Lett.* **29**, 2837–39
[5] Keldysh L V 1965 *Sov. Phys. JETP.* **20**(5) 1307-14.
[6] Normand D et al. 1990 *Opt. Lett.* **15**, 1400–02
[7] Kane D J and Trebino R 1993 *Opt. Lett.* **18**, 823–825

IOP Conf. Series: Journal of Physics: Conf. Series **1079** (2018) 012003 doi:10.1088/1742-6596/1079/1/012003

[8] Luan S, Hutchinson M H R, Smith R A and Zhou F 1993 *Meas. Sci. Technol.* **4**, 1426
[9] Franzen A, ComponentLibrary. Available at: http://www.gwoptics.org/ComponentLibrary/ Licensed under: https://creativecommons.org/licenses/by-nc/3.0/
[10] UI-5241LE. Available at: https://de.ids-imaging.com/store/ui-5241le.html. (Accessed: 21st November 2017)
[11] Wang Y, Barna S, Campbell S and Fossum E R 2001 *IEEE Workshop CCD and Advanced Image Sensors*, Lake Tahoe, Nevada, USA

IOP Conf. Series: Journal of Physics: Conf. Series **1079** (2018) 012004 doi:10.1088/1742-6596/1079/1/012004

Optimization of critical-density gas jet targets for laser ion acceleration in the collisionless shockwave acceleration regime

J L Henares[1], M Tarisien[1], P Puyuelo[1], J-R Marquès[2], T Nguyen-Bui[3], F Gobet[1], X Raymond[1], M.Versteegen[1] and F. Hannachi[1]

[1] Centre d'Etudes Nucléaires de Bordeaux Gradignan, Université de Bordeaux, CNRS-IN2P3, Route du Solarium, 33175 Gradignan, France.
[2] LULI, Ecole Polytechnique-CNRS-CEA-Université Paris VI F-91128 Palaiseau, France.
[3] CELIA, CNRS, CEA, UMR 5107, Université de Bordeaux, 33400 Talence, France.

E-mail: henares@cenbg.in2p3.fr

Abstract. Laser ion acceleration induced by high-power laser systems is nowadays an important research subject due to the large potential range of applications it could satisfy. Most of the available high-power laser facilities deliver only a few laser pulses per hour. The new facilities under development will operate at higher repetition rates (up to 10 Hz). Conventional target technologies (solid targets) and acceleration mechanisms (Target Normal Sheath Acceleration – TNSA) used so far in laser-based ion acceleration are difficult to implement at high repetition rate. New ion acceleration mechanisms such as Collisionless Shockwave Acceleration (CSA) using high density gas jets represent therefore a promising alternative. Dense gas jet targets show several advantages such as constant refresh and negligible debris production. However, full comprehension of the fluid dynamics involved in the gas jet target production is fundamental for its optimization, and at present precise data is scarce. An ongoing study of design and optimization of supersonic gas jet nozzles for laser-based ion acceleration is presented.

1. Introduction

Up to now, Target Normal Sheath Acceleration is the main laser-based mechanism used to accelerate ions from the rear side of targets (usually solid foils) [1-8]. However, other acceleration mechanisms exist which require more complex targets and laser configurations [9-13]. Most of these acceleration mechanisms produce continuous ion energy distributions. Recent studies show that an intense laser pulse can induce a collisionless shock-wave in the interaction with a critical density target [14, 15]. This shock is generated during the interaction of the high energy laser pulse with a sharp target density gradient. Ions can be accelerated by this electrostatic shock to high energies with peaked distributions (CSA regime) [16]. A gas jet nozzle at high pressure can provide the critical density needed [8, 17, 18]. It is important to note that in the experiments different mechanisms may compete and only optimization of the laser-matter interaction parameters will bring CSA to lead the acceleration process [19-20].

Recent studies show that the optimum gas jet target for a consistent pure CSA regime has to be close to the critical density or over-dense [21-23]. The critical density is defined as $n_c = \varepsilon_0 m_e \frac{\omega_0^2}{e^2}$ where ε_0 is the vacuum permittivity, m_e the electron mass, ω_0 is the angular frequency of the laser, and e is the electron charge. For a near infrared laser (1053 nm) the critical density is approximately 10^{21} cm^{-3}. This will be used as a reference in the calculations of the gas density produced by the nozzles.

Besides the density requirements, the width of the gas puff should be long enough for a proper shock-wave formation and transport in order to achieve stable acceleration. However it should not be too long to avoid depletion and slow down of the shock-wave. In addition, the gas jet shape may also play an important role and in some cases could improve the acceleration process [24]. One may conjecture that a steep slope faced to the laser beam will raise a stronger electrostatic shock-wave. On the other side of the gas jet profile, a longer relaxation will lead into a longer propagation of the shock-wave. A systematic numerical study of supersonic gas jet nozzle design is required to reach the optimal conditions for CSA mechanisms.

2. Nozzle geometry and computational fluid dynamics simulations

A convergent-divergent supersonic conical nozzle is one of the simplest geometries one can use. The simulated geometry (figure 1) consists of a reservoir where the backing pressure arrives, a convergent section, a throat, a divergent section and a chamber that mimics the experimental vacuum chamber. Systematic simulations are performed varying the backing pressure P and the nozzle parameters: diameter of the throat d, diameter of the nozzle exit D, nozzle length L and half-angle of the divergent section α, which is calculated from the previous parameters.

Computational fluid dynamics (CFD) simulations are performed using the commercial software ANSYS FLUENT [25]. An implicit density-based coupled solver (DBCS) is used with double precision accuracy green-Gauss node-based gradients of solution variables, to solve the stationary fluid flow. The standard k-omega model is used to model the turbulence. A 2D axisymmetric grid with quadrilateral cells was used, which typically consist of 1.5×10^5 cells with 2×10^{-2} average skewness ratio. The grid was adapted to the surfaces to maximize its quality, and it was checked that further refinement of the grid does not change the results.

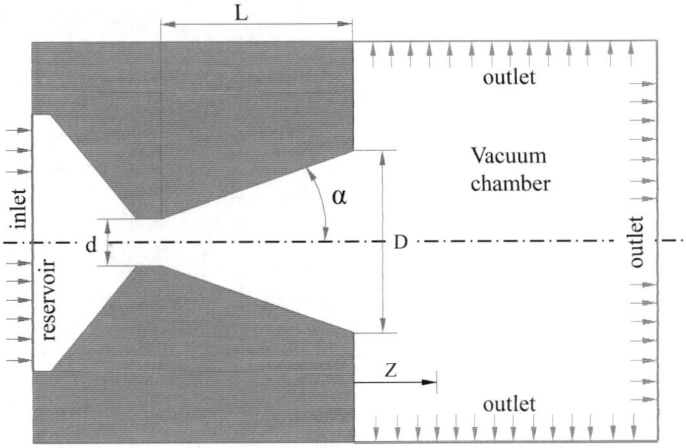

Figure 1. Scheme of the 2D axisymmetric nozzle geometry used in CFD simulations. The nozzle has a convergent section, a throat, an expansion section and exits to a chamber at vacuum. The parameters studied in this article are the inlet pressure P and the half-angle of the divergent section α, which depends on the throat diameter d, the nozzle exit diameter D and the length L. The distance from the nozzle exit is represented by Z.

The boundary conditions are: a high pressure inlet in the reservoir and three low pressure outlets in the vacuum chamber outside the nozzle. The inlet conditions are: constant laminar flux and a free expansion of the gas in the vacuum (steady solution). The medium is diatomic hydrogen, assumed to be ideal gas with a constant specific heat. The temperature of the inlet is 300 K, and the walls are supposed adiabatic. The outlet pressure was set at 10^{-5} mbar to reproduce the experimental conditions. The size of the vacuum chamber was chosen such as the gas can propagate, avoiding dilute flows. In order to verify the independence of the solution, the net imbalance of mass and energy was verified to be less than 1% of the smallest flux through the domain boundary for flux conservation.

2.1. *Supersonic flow*

The convergent-divergent geometry nozzle produces a supersonic gas flow in the diverging section. The gas flow is compressed at the throat where Mach number = 1 is reached. This flow expands and accelerates in the diverging section. The temperature and the pressure drop since the flow expansion is considered adiabatic. Our interest diverges from the classical sonic/supersonic nozzle design, where the increase of the maximum specific momentum is the main goal. In our case the objective is the increase of the density at the exit of the nozzle, while controlling other conditions such as the gas jet divergence, profile and shape. A proper understanding of the geometry of the nozzle is essential because it strongly impacts the conditions mentioned above.

Furthermore, an important feature of the supersonic propagation is the formation of Mach waves inside the nozzle and their recombination all along the flow direction. The interference of these Mach waves can increase the density in some cases and can also modify the typical Gaussian profile of the supersonic gas jet giving rise to special profiles such as a high-dense flat-top.

The evolution of the density with the backing pressure is well known to fit a linear progression [26]. Figure 2a shows the linear evolution of the density as a function of the backing pressure for three nozzle half-angles in the pressure range between 50 and 1000 bars at a distance of 500 µm from the nozzle exit. This range of pressures can be achieved with commercial gas boosters. In order to provide enough density at the nozzle exit, the working pressure has to be bigger than 600 bars. A backing pressure of 1000 bars allows to access a density close to the critical density almost independently of the nozzle parameters and over-critical densities can be reached with optimized geometries. We have chosen to work with an inlet pressure of 1000 bars and optimize the nozzles with respect to their density profile and shape.

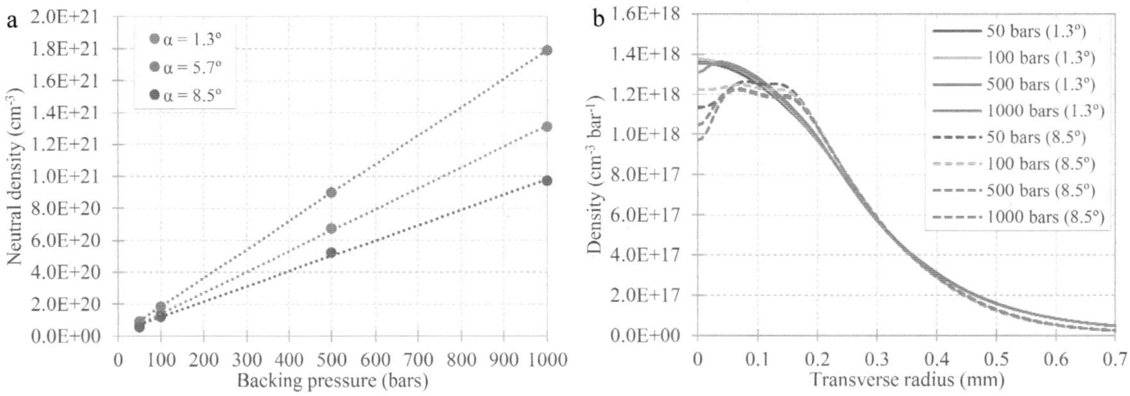

Figure 2. a) Evolution of the neutral density as a function of the backing pressure at a distance of 500 µm from the nozzle exit. Critical densities can be reached at 600 bars for some geometries but 1000 bars are needed for others. b) Evolution of the transversal density profile for a half-angle $\alpha = 1.3°$ (solid) and 8.5° (dashed) at 500 µm from the nozzle exit for different backing pressures.

As mentioned above, Mach waves created inside the diverging section can create different density profiles. In figure 2b one can observe that the density profile changes with the pressure. The difference between profiles at different pressures increased when the nozzle half-angle α was increased (dashed), however it is negligible for small half-angles (solid).

2.2. Nozzle half-angle and throat diameter

In the experiments, the laser gas jet interaction takes place at a certain distance Z from the nozzle exit (figure 1). This distance is adjusted to avoid nozzle damage due to the plasma plume expansion, while keeping the highest possible density at the laser interaction point. Systematic simulations were performed to optimize the throat diameter d and nozzle length L for sub-millimetric nozzle geometries in order to maximize the density starting from 250 µm above the nozzle exit. A density contour map was plotted for different cone half-angles at several distances from the nozzle exit. The half-angle α was varied from 0° (cylindrical case) to 10.4° by 1.3° steps. Distances Z from the nozzle exit (origin) up to 650 µm by 50 µm steps were investigated. The resulting density was expressed in n_c units (for a wavelength of 1053 nm). Figure 3 shows contour maps comparing three throat diameters d, $2/3d$ and $1/3d$. For each figure L is fixed while D is scanned.

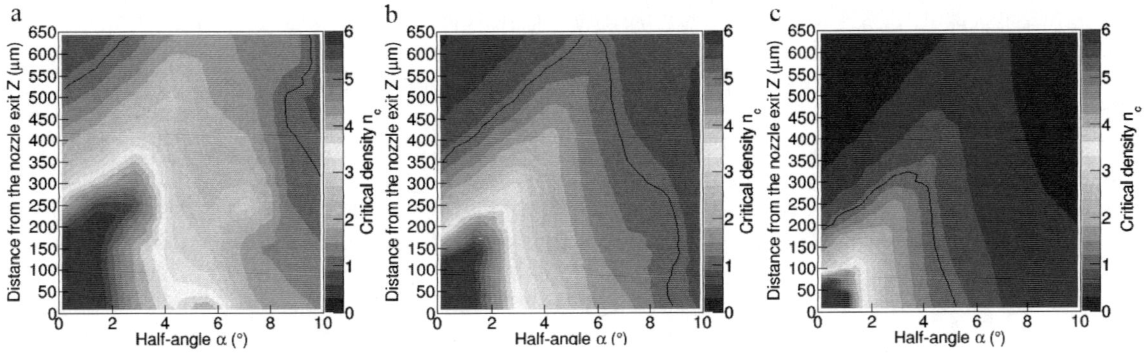

Figure 3. Interpolated contour maps of the density at 1000 bars (for a wavelength of 1053 nm) at different distances from the nozzle exit *Z* as a function of the half-angle *α* for three throat diameters (*L* is fixed): a) density-optimized nozzle throat diameter *d*, b) 2/3*d* throat diameter and c) 1/3*d* throat diameter. The position of n_c is indicated as a solid black line.

A given density can be reached using different couples of half-angle *α* and distance to the nozzle *Z*. The maximum density for one nozzle cone angle is obtained at a finite number of distances from the nozzle exit. In other words, in order to choose an optimized nozzle geometry, one should define first the interaction distance from the nozzle exit. The optimum conditions (cone half-angle and distance) can be modified in some cases by the converging Mach waves and additional density structures appear in the density contour. In figure 3a one can see the effect of these converging Mach waves at 7° and 250 µm. Figure 3b and 3c show the evolution of the density when the throat diameter is decreased. A smaller throat diameter *d* implies a slightly higher Mach number and flow velocity with a consequent decrease of density. In fact, the fundamental effect when the throat diameter is decreased is an increase of the relative size of the boundary layer thickness in the divergent section of the nozzle compared to the throat size, which leads to an increase of the turbulent kinetic energy of the flow. A relative thicker boundary layer reduces the propagation of the core flow and decreases the Mach waves rebounding off the walls [27]. This results in a drastic decrease of the density, which is a major drawback for CSA.

Figure 4 shows the contour maps for the FWHM (Full Width at Half Maximum) of the gas jet profiles performed with the same parameters as in figure 3. One can observe that a maximum density corresponds to a minimum FWHM, as a result of the converging waves. The FWHM decreases with the decrease of the gas jet throat diameter which can be advantageous to control the depletion of the laser-target interaction in CSA. However, as it was already commented, a decrease of the throat diameter can also decrease the density. A balance between these two behaviours has to be found for further optimization. Figure 4 also represents the divergence of the flow for a chosen *α* value.

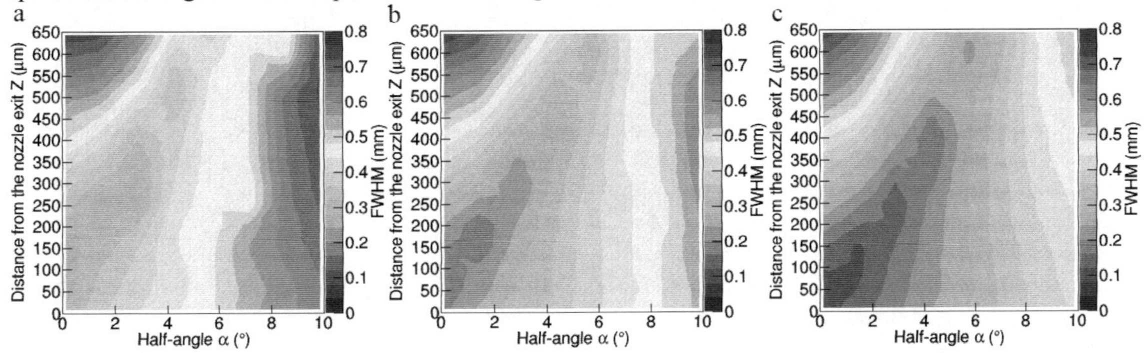

Figure 4. The FWHM interpolated contour maps at 1000 bars at different distances from the nozzle exit *Z* as a function of the half-angle *α* for three throat diameters (*L* is fixed): a) optimized throat diameter *d*, b) 2/3*d* throat diameter and c) 1/3*d* throat diameter.

IOP Conf. Series: Journal of Physics: Conf. Series **1079** (2018) 012004 doi:10.1088/1742-6596/1079/1/012004

The gas jet profile is therefore a fundamental parameter of the shock-wave formation and propagation. Some profiles can be more advantageous than others. Small angle (close to 0°) nozzles show Gaussian-like profiles. These evolve to a more pointed distributions at the minimum FWHM, and to flat-top distributions when the angle further increases (figure 5). In the contour map plot, Gaussian-like profiles are on the left border, pointed profiles correspond to minimum FWHM and flat-top profiles are on the right border. The flap-top profile can be inhomogeneous depending on how the Mach waves converge and it can present dips or wavy structures. Figure 5 shows the density profiles for a half-angle close to 0° with a Gaussian profile, pointed profile at 2.6° and flat-top profile for bigger angles.

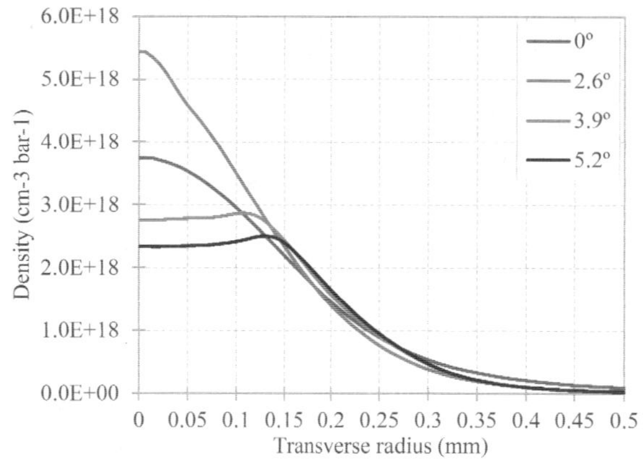

Figure 5. Transversal density profiles at 1000 bars for a half-angle α between 0° and 5.2° at 250 μm from the nozzle exit (d and L are fixed). In this case the maximum density is reached around α = 2.6° with a pointed profile. For angles smaller than this value (0°) the profiles are Gaussian-like. In the case of bigger values (3.9° and 5.2°) the profiles are flat-top.

2.3. *Nozzle length*

Figure 6 shows the change of the density contour maps for different nozzle lengths. The length does not change drastically the density, but it changes the position of the convergence of the Mach waves. A longer nozzle will dissipate more Mach waves and a shorter nozzle will not confine the flow, with a consequent decrease of density and increase of divergence.

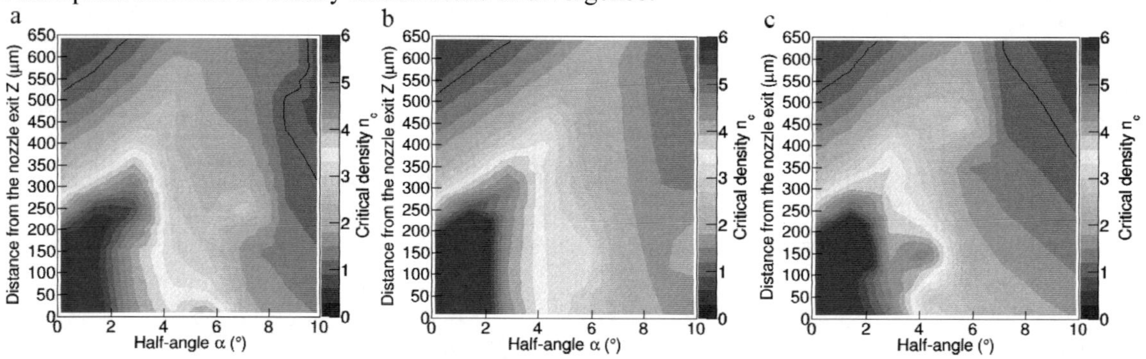

Figure 6. Interpolated contour maps of the density at 1000 bars at different distances from the nozzle exit Z as a function of the half-angle α for three lengths (d is fixed): a) density-optimized nozzle length L, b) $2L$ length and c) $0.8L$ length. The position of n_c is indicated as a solid black line.

2.4. *Reservoir and electro-valve design*

Special care has to be taken in the design of the gas transportation system (connections, tubes, valve, nozzle, etc.). An occasional formation of a supersonic flow section inside the transport system can generate instabilities in the flow that lead to turbulences and flow-blocking zones. The general rule is to avoid relatively big gas reservoir areas before the nozzle throat, use conical profiles between sections of different diameters, avoid sharp discontinuities or surface irregularities and propagate the flow from big diameters to lower diameters.

Figure 7 compares the effect of a conical transition (figure 7a) and a sharp transition (figure 7b). A sharp transition led to a perturbed flow propagation in the subsequent sections of the nozzle. In the simulation with a sharp discontinuity (figure 7b) the flow was blocked and mass conservation could not be reached due to the instabilities of the fluid flow.

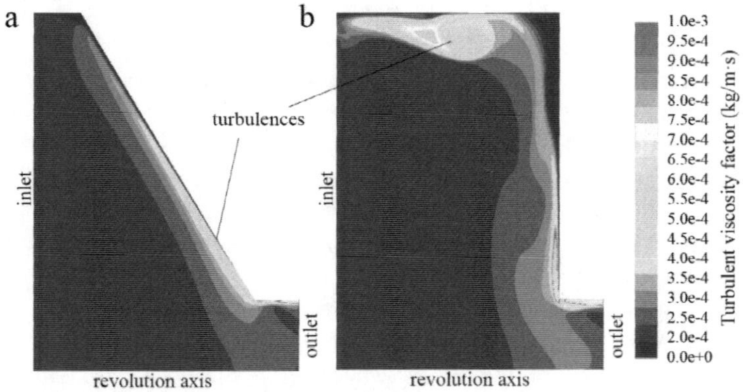

Figure 7. Effect of a sharp discontinuity in the converging region of the nozzle on the turbulent viscosity. The figures represent half of the axisymmetric view where the lower line is the revolution axis. The inlet is in the left and the outlet in the right of each figure: a) A 59° convergent cone and b) a 90° transition.

3. Conclusion

An evaluation of different conical nozzle design parameters expected to be used for laser ion acceleration in the CSA regime was performed. It was shown that CFD simulations are fundamental for a complete understanding and optimization of supersonic gas jet targets. High pressure gas jet targets offer several advantages to set up critical density targets. Small cone half-angles provided higher densities (also observed by other authors [28]) and smaller profile FWHM. The nozzle throat diameter parameter is still to be fully optimized. PIC (Particle-In-Cell) simulations will be coupled to the present study in order to investigate the relative importance between density and profile FWHM in the acceleration process.

The next step will be to compare the density and profiles obtained during this study with experimental data from interferometry density measurements [26]. Moreover a development of the simple conical nozzle to more complex designs will also require characterization. Several high-resolution tomographic techniques are currently under development that could be implemented [29-32]. It is also interesting to note the advancements in nozzle rapid machining which could ease the access to complex nozzle geometries [33].

Acknowledgments

The authors would like to thank S. Amiroudine for very fruitful discussions and comments. This work was supported by an IN2P3-CNRS grant 2016-2018.

References

[1] Wilks S C, Langdon A B, Cowan T E, Roth M, Singh M, Hatchett S, Key M H, Pennington D, MacKinnon A, and Snavely R A 2001 *Phys. Plasmas* **8** 542
[2] Clark E L, Krushelnick K, Zepf M, Beg F N, Tatarakis M, Machacek A, Santala M I K,Watts I, Norreys P A and Dangor A E 2000 *Physical Review Letters* **85** 8
[3] Maksimchuk A, Gu S, Flippo K, Umstadter D and Bychenkov V Yu 2000 *Physical Review Letters* **84** 18
[4] Snavely A, *et al.* 2000 *Physical Review Letters* **85** 2945
[5] Hatchett S P, *et al.* 2000 *Physics of Plasmas* **7** 2076
[6] Mackinnon A J, Sentoku Y, Patel P K, Price D W, Hatchett S, Key M H, Andersen C, Snavely R, and Freeman R R 2002 *Physical Review Letters* **88** 21
[7] Borghesi M 2014 *Nuclear Instruments and Methods in Physics Research Section A* **704** 6-9
[8] Prencipe I, *et al.* 2017 *High Power Laser Science and Engineering* **5** e17

IOP Conf. Series: Journal of Physics: Conf. Series **1079** (2018) 012004 doi:10.1088/1742-6596/1079/1/012004

[9] Bulanov S V, Echkina E Y, Esirkepov T Z, Inovenkov I N, Kando M, Pegoraro F, and Korn G 2010 *Physical Review Letters* **104** 135003
[10] Esirkepov T, Borghesi M, Bulanov S V, Mourou G and Tajima T 2004 *Physical Review Letters* **92** 17
[11] Wilks C, Kruer W L, Tabak M and Langdon A B 1992 *Physical Review Letters* **69** 9
[12] Macchi A, Veghini A and Pegoraro F 2009 *Physical Review Letters* **103** 085003
[13] Palmer C A J 2011 *Physical Review Letters* **106** 014801
[14] Silva L O, Marti M, Davies J R, Fonseca R A, Ren C, Tsung F S and Mori W B 2004 *Physical Review Letters* **92** 1
[15] Denavit J 1992 *Physical Review Letters* **69** 21
[16] Haberberger D, Tochitsky S, Fiuza F, Gong C, Fonseca R A, Silva L O, Mori W B and Joshi C 2011 *Nature Physics* **8** 95
[17] Sylla F, Flacco A, Kahaly S, Veltcheva M, Lifschitz A, Malka V, d'Humières E, Andriyash I and V. Tikhonchuk 2013 *Physical Review Letters* **110** 085001
[18] Musinski D L, Pattinson T R, Steinman D A, Jacobs R B 1982 *Plasma Physics* **24** 731- 742
[19] Daido H, Nishiuchi M and Pirozhkov A S 2012 *Rep. Prog. Phys.* **75** 056401
[20] Liu M-P, Xie B-S, Huang Y-S, Liu J and Yu M Y 2009 *Laser and Particle Beams* **27** 327–333
[21] d'Humières E, Lefebvre E, Gremillet L and Malka V 2005 *Physics of Plasmas* **12** 062704
[22] Stockem Novo A, Kaluza M C, Fonseca R A and Silva L O 2016 *Scientific Reports* **6** 29402
[23] Chen M, Sheng Z-M, Dong Q-L, He M-Q, Li Y-T, Bari M A and Zhang J 2007 *Physics of Plasmas* **14** 053102
[24] d'Humières E, Brantov A, Bychenkov V Yu, and Tikhonchuk V T 2013 *Phys. Plasmas* **20** 023103
[25] ANSYS FLUENT® Academic Research Mechanical, Release 18.1
[26] Malka V, Coulaud C, Geindre J P, Lopez V, Najmudin Z, Neely D, Amiranoff F 2000 *Review of Scientific Instruments* **71** 6
[27] Schmid K and Veisz L 2012 *Rev. Sci. Instrum.* **83** 053304
[28] S. Semushin, and V. Malka 2001 *Review of Scientific Instruments* **72** 2961
[29] Couperus J P, Köhler A, Wolterink T A W, Jochmann A, Zarini O, Bastiaens H M J, Boller K J, Irman A, and Schramm U 2016 *Nuclear Instruments and Methods in Physics Research A* **830** 504–509
[30] Brandi F, Giammanco F, Conti F, Sylla F, Lambert G and Gizzi L A 2016 *Review of Scientific Instruments* **87** 086103
[31] Golovin G, Banerjee S, Zhang J, Chen S, Liu C, Zhao B, Mills J, Brown K, Petersen C and Umstadter D 2015 *Applied Optics* **54** 11
[32] Tchulov O, Negro M, Stagira S, Devetta M, Vozzi C and Frumker E 2017 *Scientific Reports* **7** 6905
[33] Döpp A, Guillaume E, Thaury C, Gautier J, Phuoc K T and Malka V 2016 *Review of Scientific Instruments* **87** 073505

IOP Conf. Series: Journal of Physics: Conf. Series **1079** (2018) 012005 doi:10.1088/1742-6596/1079/1/012005

Proposed laser-targets design for space radiation simulation at ELI-NP

A Ionescu[1,2], C C Gheorghiu[1], M Bobeica[1], M Tataru[1], T Asavei[1], and V Leca[1]

[1] Extreme Light Infrastructure - Nuclear Physics, Horia Hulubei National Institute for Physics and Engineering, P.O. Box MG-6, 077125, Magurele, Romania
[2] University Politehnica of Bucharest, 313 Splaiul Independentei, Sector 6, 060042 Bucharest, Romania

aurelia.ionescu@eli-np.ro

Abstract. The space radiation environment consists of radiation fields of protons, electrons, and ions with broadband-energy spectra [1], a feature hard to mimic with classical accelerators. In this purpose, the unique parameters foreseen at the upcoming ELI-NP facility, will allow us to generate a complex, tunable, multienergetic, multicomponent, space-like radiation by means of high power lasers, in order to perform radiobiological and materials studies in a ground-based laboratory, which will improve the understanding of radiation exposure risks, and of the required radioprotection for deep space missions with human crew, and for the establishment of permanent habitats on Mars. ELI-NP's Targets Laboratory [2] has dedicated equipment for coating, (micro)assembly and characterization of, mainly, solid targets. We fabricate targets optimized for the production of cosmic-like radiation when high-power laser pulses are focused on them. Two proposed strategies for in-house target manufacturing for the above-mentioned application at ELI-NP are described: multilayer multicomponent foil and monolayer multicomponent nanostructured foil.

1. Introduction

Health effects of space radiation on astronauts is one of the major limiting factors for long-duration space missions [1], [3]. For the evaluation of radiation exposure risks, conventional accelerators (linacs and cyclotrons) are currently used to reproduce the mission-specific space radiation environment, but they are limited to the production of mono-energetic beams, while the space radiation is broadband; on the other hand, producing broadband energy radiation is the inherent regime of the laser-plasma-accelerators (LPAs). Since the space radiation features are hard to mimic with classical accelerators, LPAs can be an alternative, due to their ability to produce particle beams with the energy distribution depending on tunable laser-target parameters, and therefore they may become a viable tool for space radiation testing [4].

First reported studies on electron and proton fluxes production with energies in the range of 1-10 MeV for electrons and of 1-20 MeV for protons, similar to those from Earth's radiation belts, using laser-driven particle acceleration via Target Normal Sheath Acceleration (TNSA) mechanism [4], are promising. As for the simulation of the radiation beyond the low earth orbit (LEO), higher energies are needed to be obtained [5], [6], the solar flares providing accelerated protons, helium, and heavy ions with broad energy spectra around 100 MeV/nucleon and beyond [1].

High-Z, highly charged (Fe^{+24}, Fe^{+25}, and Fe^{+26}) ion beams ($\geq 10^6$ Fe ions per shot) with energies in the range of 0.56 - 0.89 GeV (10 - 16 MeV/n) have been successfully obtained from the interaction of a 200 TW laser with a 0.8 μm thick Al foil, covered with ~5 nm Fe impurity layer and 10 nm H_2O contaminant layer [7]. The p-polarized laser pulses had an energy of 8 J, wavelength of 0.8 μm, duration of 35 fs, a contrast of 10^{10} and a peak intensity of 10^{21} W/cm^2. PIC simulation also showed a resulting multicomponent H^+, O^{+8}, Al^{+13}, $Fe^{+18\text{-}26}$ ion beam [7].

A multicomponent H^+ and C^{+6} ion beam with energies of 12 - 93 MeV for protons, and of 36 - 174 MeV (3 - 14.5 MeV/n) for C^{6+} ions has been obtained using a PW laser system delivering circularly polarized (CP) laser pulses. An ultrathin, freestanding, 15 nm polymer (F8BT - polyfluorene derivative) target was used for the experiment, with a laser energy of 8.5 J, a duration of 30 fs, a wavelength of 800 nm, a high temporal contrast of 3 x 10^{-11}, and a maximum intensity of 7 x 10^{20} W/cm^2 [8], [9].

In view of the above, our main goal is to investigate the efficiency of novel nanostructured targets [10] with optimum properties, suitable for space-related applications. Tunable key-parameters (structure, composition, thickness, lateral dimensions, nano-pattern features, spacing, periodicity, etc.) will allow us to identify the most effective target design for which the proton and ion outputs are most appropriate, particularly for in-vitro irradiation experiments of cell monolayers [11]. In this paper we present two proposed configurations for the fabrication of targets that can be used for material testing and radiobiological studies in space-like conditions.

2. ELI-NP E5 experimental area

ELI-NP, as one of the upcoming most advanced research facilities in the world, outlined by a very high intensity laser of two 10 PW arms able to reach intensities of ~10^{22}-10^{23} W/cm^2, two 1 PW, and two 0.1 PW arms, is focused on the study of photonuclear physics and its applications. The High Power Laser System (HPLS) is based on Ti:sapphire technology, and delivers ultra-short pulses (25 fs), with a pulse repetition rate of 1/60 (10 PW), 1 (1 PW), and 10 Hz (0.1 PW), respectively [12].

The main focus of the laser driven Nuclear Physics mission of ELI-NP is the study of new nuclear phenomena induced by high power laser interaction with matter, and the development of the related applications. ELI-NP will also develop research in the field of materials behavior in extreme environments and radiobiology, with applications in shielding solutions for equipment and human crew in long term space missions [13]. This research will take place in the E5 experimental area, hosting the two 1 PW arms of the HPLS. A design of E5 experimental area is shown in figure 1.

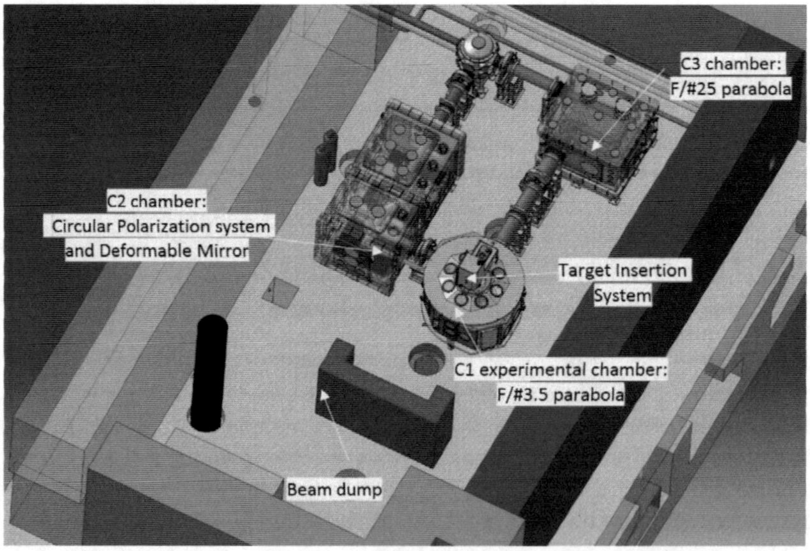

Figure 1. Design of ELI-NP E5 experimental area.

In this experimental area, some of the foreseen studies will be relevant for understanding the interaction of biological systems with tunable multicomponent radiation with broad energy spectrum.

The space radiation will be simulated by complex radiation fields generated from the interaction of the PW laser with solid targets. Particles like those from space can be obtained by tuning the laser-target interaction process, which is highly dependent of the target parameters and design. In the C1 experimental chamber (figure 1) the laser will be focused on target by means of a F/3.5 focal length off-axis parabolic mirror, employing the Target Normal Sheath Acceleration (TNSA) process.

A Circular Polarization system in the C2 chamber will make possible studies of novel acceleration mechanisms in solid targets such as Radiation Pressure Acceleration (RPA). At the same time, electron acceleration in gaseous targets will be available by focusing the 1 PW laser beam by means of a F/25 focal length off-axis parabolic mirror.

In the C1 experimental chamber, a F/3.5 parabolic mirror together with the F/25 long focusing mirror will allow the simultaneous or independent generation of ion and electron beams for material irradiation studies.

3. Proposed target strategies

In order to experimentally simulate the space radiation, consisting of protons, He and high Z and high energy (HZE) ions (C, O, Si, Fe, among others) with broad energy spectra around 100 MeV/nucleon and beyond, special targets are required. High energy protons of 93 MeV [8] have been obtained using LPAs. Aiming at the simultaneous production of HZE complex ion beam at the ELI-NP E5 experimental area, some tailored types of targets will be developed and fabricated in the ELI-NP Targets Laboratory, which is equipped with a broad range of state-of-the-art equipment for material deposition (of thick/(ultra)thin films, multi-layers, nanoparticles), processing (structuring, surface modification, patterning), surface conditioning (plasma, ion-beam, high temperature thermal treatments), micro-machining (micro-assembly, mechanical processing, cold rolling), and characterization (chemical composition, morphology/topography, structure/texture/phase). A more detailed description of the ELI-NP's Targets Laboratory is given in reference [2].

Although mono- and multi-layer, mono- and multi-component targets like 0.1 μm CH deposited layer on a 1.0 μm Au foil rear surface [14] have been reported so far in the laser-driven ion acceleration studies, they are not appropriate for this specific application, due to the unsuitable composition and energy of the accelerated ions: the ion beam contains only one or two components of interest and/or the energies of the accelerated ions are too small. For instance, C^+ ions with energies around 1 GeV (83 MeV/n) have been obtained from 200 nm-thick diamond-like carbon target [15], Fe^+ ions with energies around 16 MeV/n from 0.8 μm thick Al foil with ~5 nm Fe impurity layer and 10 nm H_2O contaminant layer [7], Si^+ ions with energies of 0.8 MeV/n from 50 nm-thick Si_3N_4 target and O^+ ions were accelerated from contaminants located on the target surface [16].

In order to overcome these issues, two novel target designs are proposed, along with a complementary study of their efficiency which will allow the identification of the most effective configuration. Accordingly, the key-parameters of the targets will be tuned (substrate material, target thickness, its lateral dimensions, nano-patterns dimensions, spacing, periodicity), with the selection of the optimum properties, for which the ion outputs will be most suitable for space applications. Preliminary fabrication and testing of simpler targets, such as a pure carbon target, an iron foil target and two components iron carbide target will also be considered for comparison purpose.

3.1. Multilayer multicomponent target

The first design considers a multilayered target consisting of one semi-conductive layer of silicon dioxide (SiO_2) followed by a layer of iron carbide (Fe_3C) on a silicon (Si) substrate with the thickness of about 350 μm. In the final step, the substrate can be completely removed, resulting in a freestanding configuration. The fabrication process is briefly presented in figure 2.

For the proposed multilayered target design, two sputtering targets (silicon dioxide, and iron carbide) will be used for the deposition of the consecutive layers. The percentage of the components will be controlled by the thicknesses of the two individual layers.

To obtain the free-standing configuration, the substrate will be removed using photolithography process (by means of mask aligner, spin coater and hot plates), and plasma etching process by means of Reactive Ion Etching (RIE) system. RIE system includes a mass spectrometer to identify the ending of the etching process, and has the capability of using both Bosch and cryogenic processes, although for this particular stage only Bosch process will be used. For the photolithography process, the wafer will be spin coated with a positive photoresist film with the thickness of approximate 8 μm.

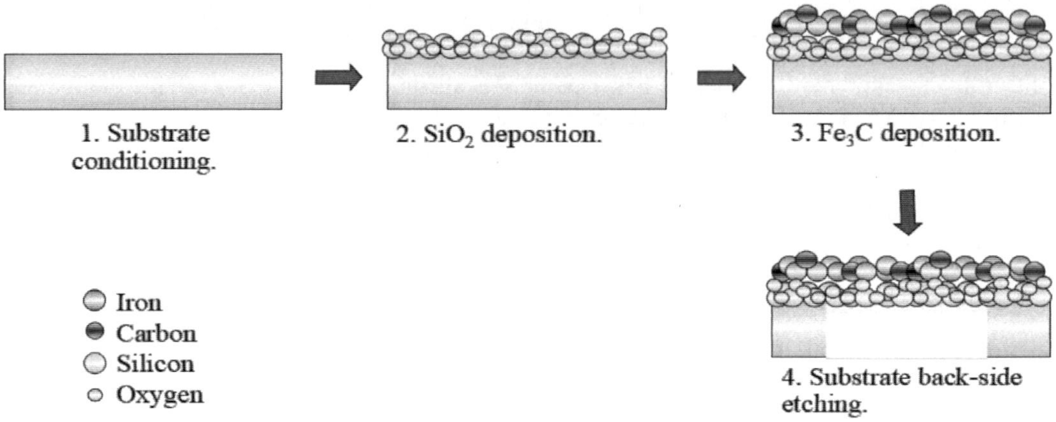

Figure 2. Multilayer target fabrication process design.

The deposition of the different layers will be performed in the Ultra-High Vacuum (UHV) Sputter Deposition cluster system (figure 3) existent within the ELI-NP Targets Laboratory. The system consists of three interconnected vacuum chambers: one vacuum chamber as a loadlock used also for the in-situ cleaning by argon ion milling of the substrate surface before the deposition, allowing a contaminant-free interface; a second vacuum chamber dedicated for the deposition of metallic and/or nitride films, and a third vacuum chamber for deposition of oxide films. The system allows for in-situ transfer of the sample between chambers, without braking the vacuum, a critical aspect for obtaining sharp interfaces between different materials of a multilayered or hybrid (e.g., metal/oxide) structure.

The lateral sizes of the free-standing film will be about 200 μm x 200 μm and the total thickness will range from 200 nm down to few tens of nanometers. Also, the order and the thickness of the individual layers can be varied in the total thickness range, to identify the optimal configuration of the target.

The main advantages of the proposed targets, with a multilayer configuration, is that it allows us to obtain a specific multicomponent structure in a relatively simple fabrication process, and to cover a broader range of thicknesses. The group of D. C. Carroll reported that for carbon ions acceleration, the energy was decreased by using a multicomponent target (CH), in comparison with the pure carbon target, for a given target thickness [14], which can be a disadvantage that needs to be eliminated.

3.2. Monolayer multicomponent nanostructured target

The second proposed design is a monolayer, multicomponent foil. The main feature of this strategy is the target fabrication which uses one multicomponent sputtering target, giving the possibility of obtaining a homogeneous multi-species single layer target type. In this approach, the target thickness can be significantly reduced to few nanometers (ultra-thin films), which is more difficult to obtain in the multilayered target case.

Figure 3. ELI-NP Targets Laboratory UHV system [17].

Moreover, in order to enhance the laser energy absorption, iron nanoparticles deposition can be performed in an additional step. The multicomponent nanostructured target fabrication process is presented in figure 4.

For the fabrication of the second design, a multicomponent, homogeneous sputtering target will be manufactured, consisting of a mixture of different elements (Si, Fe, C, O) in the desired stoichiometry, prepared as a pellet that will be used as sputtering target for the UHV system. The lateral size and the thickness range will be similar to those for the multilayer targets. The freestanding target structure will be obtained similarly with the previous case, by back-side plasma etching of the substrate (Si wafers), while the metallic nanoparticles deposition will be performed using NANOGEN tool from the UHV sputtering system.

The presence of the nanoparticles on the target front-side is expected to enhance the absorption of the laser beam, and consequently, to increase the maximum ion beam energy and to improve the energy conversion efficiency from laser to the ion beam.

The production of the homogeneous multicomponent monolayer target, comparing to that of the multilayer target, is a more complex process, as it requires additional manufacturing steps, such as fabrication of the pellet used as sputtering target.

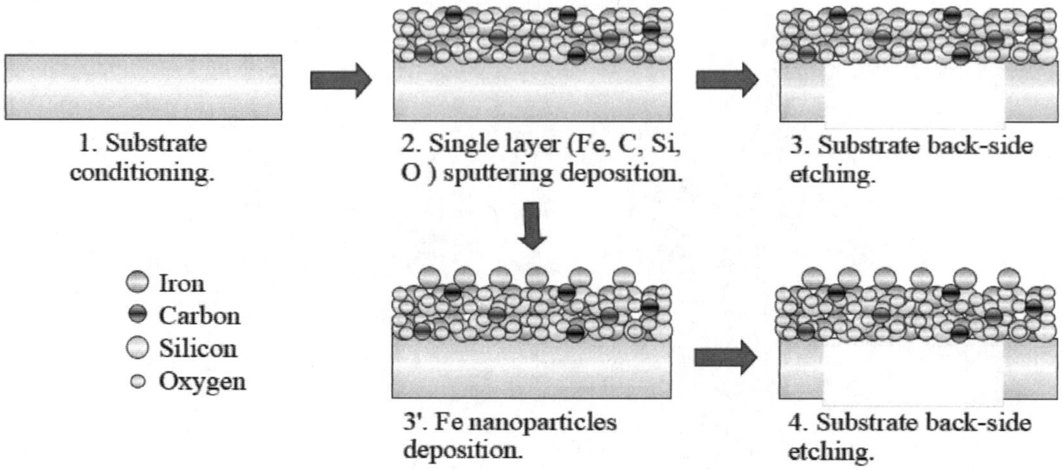

Figure 4. Monolayer multicomponent nanostructured target fabrication process design.

The water layer usually present on the target surface will be removed by UV heating of the target, or by other appropriate method, and this is expected to result in an increase in the energy of the heavier ions [18].

4. Conclusions

Two proposed multicomponent target designs were presented, a multilayered structure and an ultra-thin single-layered target. These targets will be used for the simultaneous production of a multicomponent beam with composition similar to that of space radiation, therefore recreating some of the conditions from space in a ground-based laboratory, by laser-driven mixed ion species acceleration. The novelty in concept of the two proposed target configurations consists in four high-Z specific elements (C, Fe, O, Si) in the target composition.

Furthermore, PIC simulations of the experiment with the proposed targets are currently underway, taking into consideration different target thicknesses (from a few nm to about 1 μm), in order to identify and fabricate the most suitable targets for laser-driven accelerated ion beams with space-like radiation features: complex radiation with broad energy spectra around 100 MeV/nucleon. For both proposed designs the final range of target thicknesses will be chosen with respect to the laser parameters, in order to maximize the energy of the accelerated ions.

The nanostructured targets with iron nanoparticles deposited on the laser side of the target is expected to increase the laser energy absorption, the laser energy coupling to the plasma, and consequently, the ion acceleration process. Moreover, the targets will be tested at ELI-NP PW laser facility for efficiency studies and performance analysis.

Acknowledgments

The authors gratefully acknowledge the support from ELI-NP Phase II, a project co-financed by the Romanian Government and European Union through the European Regional Development Fund - the Competitiveness Operational Programme (1/07.07.2016, COP, ID 1334).

References

[1] Norbury J W et al. 2016 Galactic cosmic ray simulation at the NASA Space Radiation Laboratory *Life Sci. Space Res.* **8** 38-51

[2] Gheorghiu C C, Leca V, Popa D, Cernaianu M O and Stutman D 2016 *J. Instrum.* **11** C10011

[3] Hidding B, Karger O, Königstein T, Pretzler G and Rosenzweig J B 2013 Study of Space Radiation Effects with Laser-Plasma-Accelerators *ESA NPI Activity Final Report* 4000102854

[4] Hidding B et al. 2017 Laser-plasma-based Space Radiation Reproduction in the Laboratory *Scientific Reports* 7 42354
[5] Mewaldt R A, Davisa A J, Binnsb W R, De Nolfoc G A, Georged J S, Israelb M H, Leskea R A, Stonea E C, Wiedenbecke M E and Von Rosenvingec T T 2005 *29th Int. Cosmic Ray Conf. (Pune)* 00 pp 101-104
[6] Schimmerling W 2016 *Life Sciences in Space Research* **9** 2-11
[7] Nishiuchi M et al. 2015 *Physics of Plasmas* **22** 033107
[8] Kim I J et al. 2016 *Physics of Plasma* **23** 070701
[9] Choi I W et al. 2011 *Applied Physics Letters* **99** 181501
[10] Margarone D et al. 2013 *Proc. of SPIE* 8780 878023
[11] Bobeica M, Aogaki S, Asavei T, Cernaianu M O, Ghenuche P, Negoita F and Stutman D 2016 Radiobiology Experiment Design and Modeling for Space Applications at ELI-NP *Proc. Int. Conf. of Aerospace Sciences "AEROSPATIAL 2016"* (Bucharest) (Bucharest: Incas) p 1
[12] Negoita F et al. 2016 *Romanian Reports in Physics* **68**, Supplement, S37–S144
[13] Asavei T et al. 2016 *Romanian Reports in Physics* **68**, Supplement, S275–S347
[14] Carroll D C et al. 2010 New Journal of Physics **12** 045020
[15] Jung D et al. 2013 Physics of Plasmas **20** 083103
[16] McGuffey C et al. 2016 New Journal of Physics **18** 113032
[17] Design and drawing by Mantis Depositions Ltd, UK: http://www.mantisdeposition.com
[18] McKenna P et al. 2004 Physical Review E **70** 036405

IOP Conf. Series: Journal of Physics: Conf. Series **1079** (2018) 012006 doi:10.1088/1742-6596/1079/1/012006

Measurement of the particle number density in a pulsed flow gas cell with a second-harmonic interferometer

F Brandi[1,2], P Marsili[3], F Giammanco[3], F Sylla[4] and LA Gizzi[1]

[1] Intense Laser Irradiation Laboratory (ILIL), Istituto Nazionale di Ottica (CNR-INO), Via Moruzzi 1, 56124 Pisa, Italy

[2] Istituto Italiano di Tecnologia (IIT), Via Morego 30, 16163 Genova, Italy

[3] Dipartimento di Fisica, Università degli Studi di Pisa, Largo B. Pontecorvo 3, 56127 Pisa, Italy

[4] SourceLAB SAS, 86 rue de Paris, 91400 Orsay, France

E-mail: fernando.brandi@ino.cnr.it

Abstract.
A high-sensitivity high-speed second-harmonic interferometer is used to monitor the particle number density inside a pulsed flow gas cell designed for laser wakefield acceleration. The interferometer can precisely follow the particle density temporal evolution therefore offering a practical way to control in real-time the target density during laser-plasma interaction. The presented results are relevant for the evaluation of density diagnostic tools for flow gas cells used as laser-plasma acceleration stages.

1. Introduction

The demonstration of laser wakefield acceleration (LWFA) of electrons to the GeV level [1, 2] opens the way for the application of LWFA-based accelerators and high photon energy radiation sources [3] within user oriented facilities with superior beam quality and reliability necessary for actual high-level applications, as envisaged within the EuPRAXIA project [4]. The transition from fundamental research to actual implementation must go along with a full control and tailoring of the laser-plasma based accelerator, and specifically of the plasma free-electron density. In LWFA the plasma in the gas target is created either by the laser pulse itself or pre-generated by an electric discharge or by a second laser pulse.

Gaseous targets used in LWFA are: *i) supersonic gas jets* that are easy to implement, allow for a good control over the peak particle number density [5], but produce density profiles that can vary shot-to-shot due to reproducibility of valve operation over time and turbulent flow, and prevent high repetition rate operation due to pulsing capability. *ii) Capillary discharges* that provides guiding of the laser up to centimeters, thus increasing the acceleration length [1], but can get damaged during usage and may require somehow sophisticated density diagnostics methods [6] which pose technical challenges; *iii) flow gas cells* that are very good candidates to avoid the above mentioned limitations, allowing for a stable and controllable laser-plasma interaction even at high repetition rate along with easily tunable accelerator length [7–14]. Flow gas cells are also suitable to be implemented in multi-stage accelerators [15, 16], which

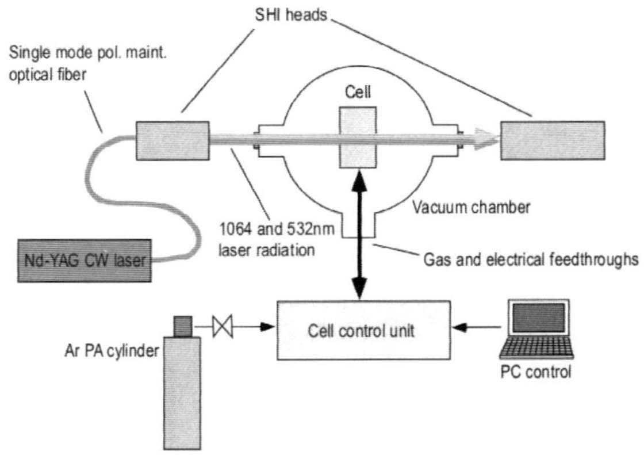

Figure 1. Schematic of the experimental apparatus.

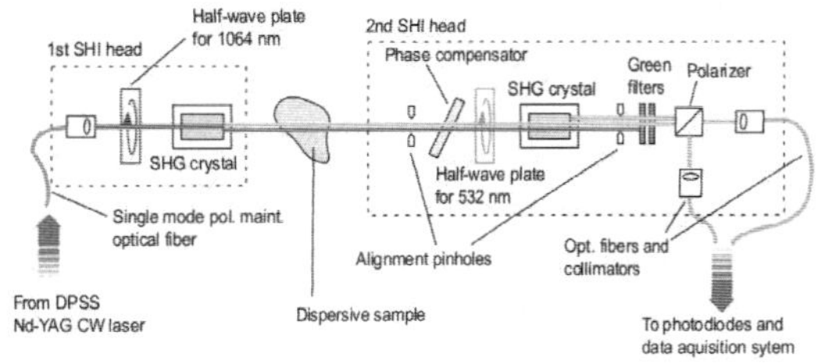

Figure 2. Schematic of the second-harmonic interferometer.

is important in the perspective of designing and implementing LWFA-based facilities. In this context the reliable and robust measurement and control of the free-electron density is a crucial aspect, and interferometry plays a major role as diagnostic tool being non-invasive and versatile. Typical two-arm interferometers [17] suffer from a high sensitivity to environmental conditions which limits their use outside the laboratory environment. Nomarski and folded two-arm interferometers provide better stability compared to standard two-arm interferometers, are widely used in research laboratories [18–21], and provide an interferogram from which the phase is retrieved applying image analysis software and phase-unwrapping algorithms. The lengthy data analysis limit their use in real-time measurements, while they may be suitable for off-line measurements during alignment and tuning of the laser-plasma accelerator stages.

When stability and ease of implementation matters, an alternative robust interferometric method is the second-harmonic interferometer (SHI), also called dispersion interferometer [22]. The SHI is a single-arm, two-color interferometer, which is sensitive to the change of refractive index between the fundamental and second-harmonic wavelength, measured phase shift given by $\Delta\phi = \frac{4\pi}{\lambda}\int_L \Delta n(\lambda) dl = \frac{4\pi}{\lambda} L\overline{\Delta n}(\lambda)$, where $\Delta n(\lambda) = n(\lambda) - n(\lambda/2)$, $n(\lambda)$ is the refractive index, λ is the wavelength, and L the optical path in the sample. Being a fully common-path interferometer the SHI is insensitive to vibration when compared with typical two-arm interferometers, which allows long term stable operation even in a harsh environment [23].

In this work, the latest results on the use of a high-speed ($\sim \mu$s) and high-sensitivity ($\sim$ mrad) SHI based on a CW Nd:YAG laser to measure the particle number density inside a pulsed flow gas cell in vacuum are presented.

2. Experiment

Fig.1 shows a schematic block diagram of the experimental apparatus. The cell (model SL-ALC, SourceLAB) is placed in a cylindrical vacuum chamber, evacuated by a turbo-molecular pump to 10^{-5} mbar. The gas cell comprises two 600 μm-diameter apertures to allow the gas to flow in the vacuum chamber, two lateral glass windows at $L = 35$ mm to enable transverse interferometry. When the gas pulse flows out of the cell the pressure in the chamber increases shortly, not exceeding however values of the order of 10^{-2} mbar. The vacuum chamber is equipped with three KF50 vacuum flanges on the lateral surface. Two lie opposite along a diameter and are used to mount BK7 optical precision windows, anti-reflection coated for 1064 nm and 532 nm wavelengths, providing entrance and exit of the interferometer beams. The third flange hosts the gas and electrical feedthroughs between the cell and the controller. The argon gas at 2

bar enters the control unit which then sends a gas puff to the cell at a pre-set pressure and duration. It is noted that if the cell would be continuously feed with gas or long gas pulses were used, the effect from back-ground density build up in the chamber would become visible during the interferometric measurement. Such effect was not revealed for the gas pressure and pulse length used. In case longer pulses and/or higher pressures would be needed the vacuum system (chamber plus pump) would have to be adequately scaled in order to avoid density build up in the interaction chamber.
Fig. 2 shows a detailed sketch of the fully fiber-coupled SHI interferometer used in the experiment [24–26], which is here briefly described. The radiation of a CW DPSS Nd-YAG laser is sent to the first interferometer head using a single mode polarization maintaining optical fiber. The laser light emerging from the fiber is collimated to a diameter of 1 mm. A half-wave plate is used to adjusts the polarization, in order to ensure an optimal harmonic generation in the first type-I SHG crystal. Both the 1064 nm and 532 nm beams leave the first head and are sent collinearly thought the cell. In the cell the two components suffer a de-phasing due to the gas dispersion and proportional to the gas number density. The transmitted beams enter the second head where a half-wave plate rotates the second-harmonic polarization by an angle of 90° leaving the fundamental beam polarization unchanged. A tilted glass slab adds a controlled de-phasing acting as a compensator, in order to tune the phase difference to an optimal working point. In a second type-I SHG crystal the 1064 nm beam is duplicated again. Following the filtering out of the residual 1064 nm beam, the two 532 nm beams with crossed polarizations enters a polarizing cube oriented by an angle of 45°, where the two beams are mixed giving rise to two complementary interference patterns. The beams emerging from the beam-splitter cube are finally collected by two fiber optic cables and sent to photodiodes, directly connected to ultra low-noise transimpedance amplifiers. The output signals are acquired by a 15 MHz USB digitizing oscilloscope controlled by a labview software which calculates in real-time the ratio between the difference and the sum of the digitalized signals. The recorded quantity is equal to $V\sin(\Delta\phi + \phi_0) + \alpha$ [24], where $\phi_0 \ll 1$ is the off-set phase that can be controlled acting on the phase compensator, V is the fringe visibility, and $\alpha \ll 1$ is related to the detector responsivities. The visibility is directly obtained by scanning the phase compensator over half-fringe [25, 27] and it is $V = 0.9$.

3. Results

The Gladstone-Dale relation between the refractive index n and the number density N, i.e., $(n-1) \propto N$, is used to obtain the particle number density from the measured phase by the equation $\overline{N} = \frac{\lambda}{4\pi L}\frac{N_0}{\Delta n_0}\Delta\phi = 1.63\Delta\phi \times 10^{19}$ cm^{-3}, where $N_0 = 2.69 \times 10^{19}$ cm^{-3} is the Loschmidt constant and $\Delta n_0 = 4 \times 10^{-6}$ the difference of the refractive index of argon at 1064 nm and 532 nm [28].
The results of the systematic measurements for a 100 ms gas pulse at various values of the pre-set backing pressure are reported in Fig. 3. The zero point represent the time when the trigger is sent to the cell's controller. In less then 100 ms from the trigger the filling up of the cell starts and lasts for about 100 ms. Then the gas density in the cell drops exponentially with a decay time of ~0.9 s.
In Fig. 4 a comparison between 100 ms and 500 ms long gas pulses is shown for two values of the backing pressure. As expected the filling up of the cell lasts longer for the longer gas pulse and the achieved peak density value is larger by a factor ~1.5.
Fig. 5 shows the values of the peak density value obtained in the experimental conditions investigated, while the dashed-line indicate the density estimated from the ideal gas law at the preset backing pressure and ambient temperature, reported as reference.

IOP Conf. Series: Journal of Physics: Conf. Series **1079** (2018) 012006 doi:10.1088/1742-6596/1079/1/012006

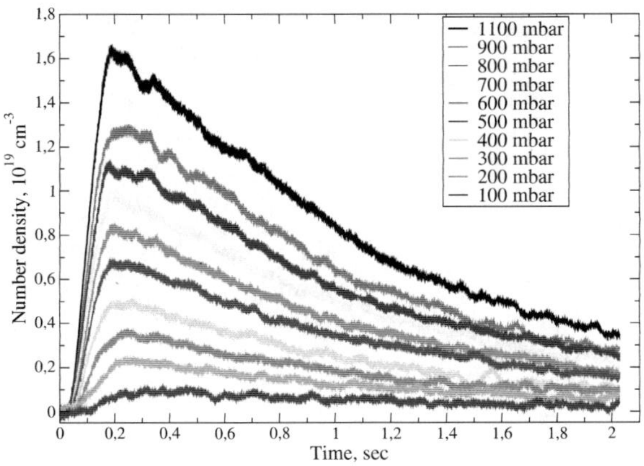

Figure 3. Time evolution of the average particle number density for a 100 ms gas pulse at various backing pressure settings.

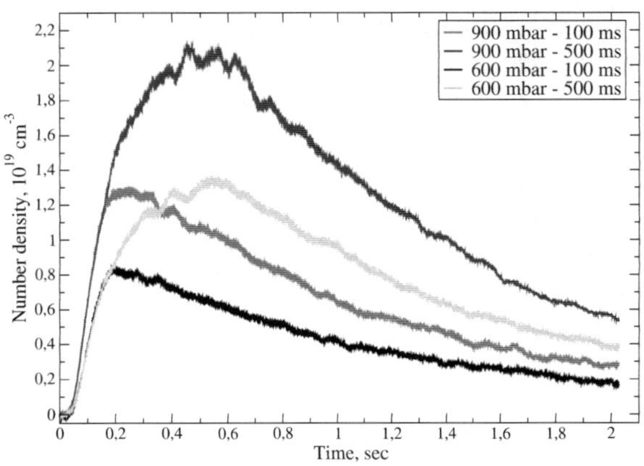

Figure 4. Time evolution of the average particle number density for a 100 ms and 500 ms gas pulse at two backing pressure settings.

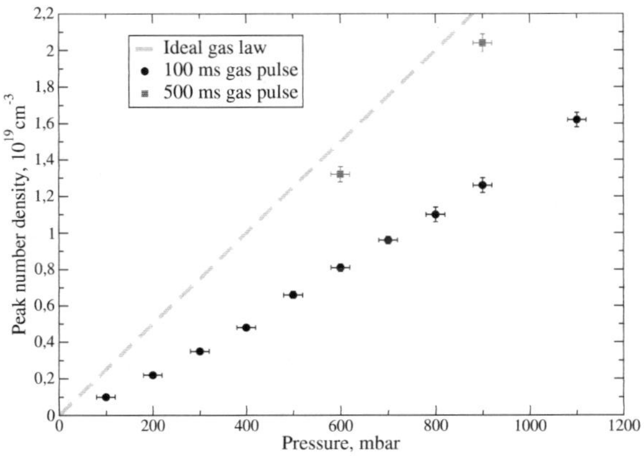

Figure 5. Peak value of the particle number density for 100 ms and 500 ms gas pulse at various backing pressure settings.

4. Conclusions

A second-harmonic interferometer is used to monitor in real-time the particle number density in a commercial pulsed flow gas cell designed for LWFA. It is found that diagnostic method is successful in measuring the density temporal evolution inside the cell from the filling up to the evacuation in the range up to 10^{19} cm^{-3}. The achieved peak value is less then what estimated from the ideal gas law especially for the shortest gas pulse, therefore characterization and continuous monitoring of the gas density is necessary in order to tune and control the laser-plasma interaction process inside the cell. The experiment performed and the results obtained are important towards the implementation of gas cells as laser-plasma based acceleration stages in a fully controlled and user oriented particle accelerator facility, as that envisaged within the EuPRAXIA design study project.

It is noted that the SHI presented can be used also to measure the particle density for other gases typically used in LWFA, like H_2 and He. The difference between the refractive indices at 1064 nm and 532 nm in hydrogen and helium at STP are 2.8×10^{-6} and 2.2×10^{-7} respectively [28]. Therefore, the measurement can be performed with hydrogen instead of argon. In case of helium, the expected phase shift at STP is ~90 mrad, therefore the SHI with a noise level down to less the 1 mrad [24] is capable of measuring He number density within the cell in the order of 10^{17} cm^{-3}.

In perspective, the development of a 2D imaging version of the SHI would allow to monitor

IOP Conf. Series: Journal of Physics: Conf. Series **1079** (2018) 012006 doi:10.1088/1742-6596/1079/1/012006

samples with non-uniform and/or non-cylindrically symmetric spatial density distribution, like gas jets from square nozzles and/or with shock fronts. There are reports in the literature about 2D second-harmonic interferometry [29–31] however more research and development activity is necessary to validate the imaging SHI methodology as viable diagnostics for LWFA gas target.

Acknowledgments

FB and LG acknowledge the financial support from the European Unions Horizon 2020 research and innovation programme Grant Agreement No 653782 EuPRAXIA, and from the Istituto Italiano di Tecnologia (convenzione operativa IIT CNR-INO prot. n. 0010983, 26/11/2013).

References

[1] Leemans W P *et al.* 2014 Multi-GeV Electron Beams from Capillary-Discharge-Guided Subpetawatt Laser Pulses in the Self-Trapping Regime *Phys. Rev. Lett.* **113** 245002

[2] Hyung T K *et al.* 2017 Stable multi-GeV electron accelerator driven by waveformcontrolled PW laser pulses *Sci. Rep.* **7** 10203

[3] Gizzi L A *et al.* 2009 An integrated approach to ultraintense laser sciences: The PLASMON-X project *Eur. Phys. J. Spec. Top.* **175** 3-10

[4] Walker P A *et al.* 2017 Horizon 2020 EuPRAXIA design study *J. Phys.: Conf. Ser.* **874** 012029

[5] Brandi F and Giammanco F 2011 Temporal and spatial characterization of a pulsed gas jet by a compact high-speed high-sensitivity second-harmonic interferometer *Opt. Express* **19** 25479–87

[6] Daniels J, Tilborg J van, Gonsalves A J, Schroeder C B, Benedetti C, Esarey E and Leemans W P 2015 Plasma density diagnostic for capillary-discharge based plasma channels *Phys. Plasmas* **22** 073112

[7] Osterhoff J *et al.* 2008 Generation of Stable, Low-Divergence Electron Beams by Laser-Wakefield Acceleration in a Steady-State-Flow Gas Cell *Phys. Rev. Lett.* **101** 085002

[8] Corde S *et al.* 2013 Observation of longitudinal and transverse self-injections in laser-plasma accelerators *Nature Comm.* **4** 1501

[9] Aniculaesei C, Kim H T, Yoo B J, Oh K H and Nam C H 2018 Novel gas target for Laser Wakefield Accelerators *Rev. Scient. Intrum.* **89** 025110

[10] Kononenco O *et al.* 2016 2D hydrodynamic simulations of a variable length gas target for density down-ramp injection of electrons into a laser wake field accelerator *Nucl. Instr. Meth. Phys. A* **829** 125-9

[11] Audet T L *et al.* 2016 Electron injector for compact staged high energy accelerator *Nucl. Instr. Meth. Phys. A* **829** 304-8

[12] Lee P, Maynard G, Audet T L, Cros B, Lehe R and Vay J-L 2016 Electron injector for compact staged high energy accelerator*Phys. Rev. Accel. Beams* **19** 112802

[13] Audet T L *et al.* 2016 Investigation of ionization-induced electron injection in a wakefield driven by laser inside a gas cell *Phys. Plasmas* **23** 023110

[14] Vargas M *et al.* 2014 Improvements to laser wakefield accelerated electron beam stability, divergence, and energy spread using three-dimensional printed two-stage gas cell targets *Appl. Phys. Lett.* **104** 174103

[15] Pollock B B *et al.* 2011 Demonstration of a narrow Energy Spread, ∼ 0.5 GeV Electron Beam from a Two-stage Laser Wakefield Accelerator *Phys. Rev. Lett.* **107** 045001

[16] Liu J S *et al.* 2011 All-Optical Cascaded Laser Wakefield Accelerator Using Ionization-Induced Injection *Phys. Rev. Lett.* **107** 035001

[17] Gizzi L A *et al.* 2006 Femtosecond interferometry of propagation of a laminar ionization front in a gas *Phys. rev. E* **74** 036403

[18] Feister S, Nees J A, Morrison J T, Frische K D, Orban C, Chowdhury E A and Roquemore W M 2014 A novel femtosecond-gated, high-resolution, frequency-shifted shearing interferometry technique for probing pre-plasma expansion in ultra-intense laser experiments *Rev. Scient. Intrum.* **85** 11D602

[19] Park J, Baldis H A and Chen H 2016 The implementation and data analysis of an interferometer for intense short pulse laser experiments *High Power Laser Science and Engineering* **4** E26, doi:10.1017/hpl.2016.21

[20] Kalal M, Slezak O, Martinkova M and Rhee Y J 2010 Compact Design of a Nomarski Interferometer and Its Application in the Diagnostics of Coulomb Explosions of Deuterium Clusters *Journal of the Korean Physical Society* **56** 287-94

[21] Ruiz-Camacho J, Beg F N and Lee P 2007 Comparison of sensitivities of Moire deflectometry and interferometry to measure electron densities in z-pinch plasmas. *J. Phys. D: Appl. Phys.* **40** 2026-32

[22] Hopf F, Tomita A, and Al-Jumaily G 1980 Second-harmonic interferometers *Opt. Lett.* **5**, 386–8

[23] Brandi F, Giammanco F, Harris W S, Roche T, Trask E and Wessel F J 2009 Electron density measurements of a field-reversed configuration plasma using a novel compact ultrastable second-harmonic interferometer *Rev. Sci. Instrum.* **80** 113501

[24] Brandi F and Giammanco F 2007 Versatile second-harmonic interferometer with high temporal resolution and high sensitivity based on a continuous-wave Nd:YAG laser *Opt. Lett.* **32** 2327–29

[25] Brandi F and Giammanco F, 2008 Harmonic interferometry in the visible and UV, based on second- and third-harmonic generation of a 25 ps mode-locked Nd:YAG laser *Opt. Lett.* **33** 2071-73

[26] Brandi F, Giammanco F, Conti F, Sylla F, Lambert G and Gizzi LA 2016 Note: Real-time monitoring via second-harmonic interferometry of a flow gas cell for laser wakefield acceleration *Rev. Sci. Instrum.* **87** (8) 086103

[27] Brandi F, Marsili P and Giammanco F 2008 Compact high-speed high-sensitivity second-harmonic interferometer for electron density measurement in AIP conference proceedings: Burning plasma diagnostics, **988**, 132–5

[28] Velsko S P and Eimerl D 1986 Precise measurements of optical dispersion using a new interferometeric technique *Appl. Opt.* **25**, 1344–49

[29] Minoshirna K and Matsumoto H 1997 In-situ measurements of shapes and thicknesses of optical parts by femtosecond two-colour interferometry *Opt. Commun.* **138** 6-10

[30] Abraham E, Minoshima K and Matsumoto H 2000 Femtosecond laser-induced breakdown in water: time-resolved shadow imaging and two-color interferometric imaging *Opt. Commun.* **176** 441-452

[31] Jobes, F C and Bretz N L 1997 A prototype imaging second harmonic interferometer. *Rev. Scient. Intrum.* **68** 709

IOP Conf. Series: Journal of Physics: Conf. Series **1079** (2018) 012007 doi:10.1088/1742-6596/1079/1/012007

Wafer-scale fabrication of target arrays for stable generation of proton beams by laser-plasma interaction.

R L Zaffino[1], M Seimetz[2], A Ruiz[3], I Sanchez[3], P Mur[2], P Bellido[2], R Lera[3], D Quirion[1], L Martín[4], J Benlliure[4], J M Benlloch[2], M Lozano[1] and G Pellegrini[1]

[1]Instituto de Microelectrónica de Barcelona IMB-CNM-CSIC 08193 Barcelona, Spain
[2]Instituto de Instrumentación para Imagen Molecular I3M-CSIC, Universitat Politécnica de València, Camino de Vera s/n, Ed. 8B-N-1[a], 46022 Valencia, Spain
[3]Proton Laser Applications S. L., Av. Vilafranca del Penedés 11a, 08734 Olérdola, Barcelona, Spain
[4]Dto Fisica de Particulas, Campus Vida, USC, Rua Xoaquín Días de Rábago S/N, 15782 Santiago de Compostela, Spain

rossella.zaffino@imb-cnm.csic.es

Abstract. Large-scale fabrication of targets for laser-driven acceleration of ion beams is a prerequisite to establish suitable applications, and to keep up with the challenge of increasing repetition rate of currently available high-power lasers. Here we present manufacturing and test results of large arrays of solid targets for TNSA laser-driven ion acceleration. By applying micro-electro-mechanical-system (MEMS) based methods allowing for parallel processing of thousands of targets on a single Si wafer, sub-micrometric, thin-layer metallic membranes were fabricated by combining photolithography, physical and chemical vapor deposition, selective etching, and Si micromachining. These structures were characterized by using optical and atomic force microscopy. Their performance for the production of laser-driven proton beams was tested on a purpose-made table-top Ti:Sapphire laser system running at 3 TW peak power with a contrast over ASE of 10^8. We have performed several test series achieving maximum proton energy values around 2 MeV.

1. Introduction

In the next years a series of high power laser installations will be available worldwide with the ability of operating at repetition rates (1÷10 Hz) never before experienced. This working regime will require thousands of consumable targets per day to sustain experimental campaigns, so that it is extremely relevant to develop methods allowing for mass production and characterization of targets with high versatility for the different experimental needs [1]. In addition to production, target positioning, alignment, and refreshment are of the same importance for the proper target design of a given application. Targets for high-power laser experiments are of different densities ranging from gaseous to solid state, including foams and cryogenic materials. Here, we restrict our attention to solid state

IOP Conf. Series: Journal of Physics: Conf. Series **1079** (2018) 012007 doi:10.1088/1742-6596/1079/1/012007

targets for laser-driven proton acceleration through Target Normal Sheath Acceleration (TNSA) [2] which is the prevailing mechanism in the experiments conducted so far. Targets used for TNSA laser irradiation have been typically thin foils of metallic or dielectric materials with thicknesses up to tens of microns [3]. More recently, the improvement in laser contrast over ASE, and the advancement in material science technologies, allowed for exploring sub-micrometric thicknesses, with promising results with respect to the maximum proton energies. In this work, we made use of MEMS based manufacturing to fabricate thin membranes of conductive materials integrated in a Si wafer. These were used as targets for TNSA laser irradiation with a purpose-made 3 TW laser system. MEMS processing such as pattern transfer, material deposition and selective etching make it possible to achieve simple (2D) or complex (3D) target configurations, allowing for parallel manufacturing of thousands of targets with high reproducibility of specimen belonging to the same wafer, and between wafers of the same batch. For these reasons, and due to their availability on industrial scale developed by the semiconductor industry, they appear especially adequate for the technological developments expected within the field of high-power lasers.

2. Target fabrication and characterization.

A schematic of the process flow applied to obtain a single membrane is shown in Fig.1 joint with an example of the pattern transferred onto a standard p-type <100> silicon wafer, double side polished, with a thickness of 525 μm. The principal steps are: photolithography on wafer front side, followed by back side metallization, and metal etching. For some samples, we also added a back side photolithography used to reduce the metallic pattern corresponding to the window opened on the front wafer side.

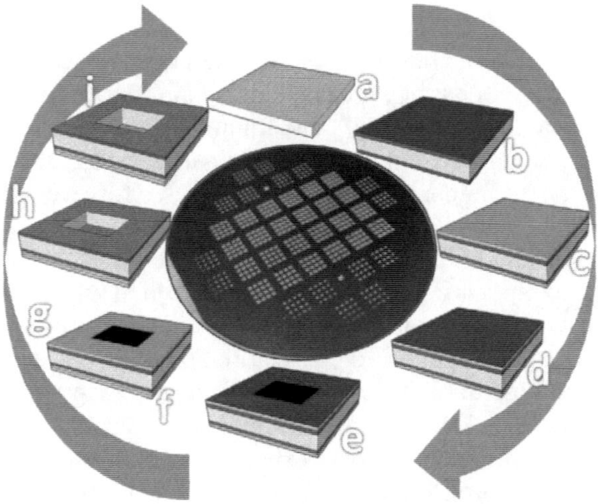

Figure 1. Schematic of the process flow for the fabrication of an individual membrane. In the centre an example of a 4 inch wafer with embedded membranes.

As first step Si wafers (Fig.1-a) were wet oxidized to grow either a 100 nm (at 950°C) or 400 nm thick (at 1100°C) SiO_2 layer, Fig. 1-b. This was followed by the deposition of a 180 nm thick low pressure chemical vapour deposition (LPCVD) layer of Si_3N_4 (Fig.1-c) on front side of the wafer corresponding to the laser entry side of the target. In the subsequent step, a 1.2 μm thick layer of the (positive) photoresist HIPR 6512 (Figl-d) was spun on the same wafer side and exposed to UV light (using a Canon PLA600 mask aligner) through a photomask to define the area and the spatial distribution of the target membranes (Fig.1-e-f). Thereafter, etching of exposed Si_3N_4/SiO_2 layers was achieved by a dry process using a mixture of $Ar/CHF_3/CHF_4$ (in the Applied Material Precision 5000 Mark II P5

IOP Conf. Series: Journal of Physics: Conf. Series **1079** (2018) 012007 doi:10.1088/1742-6596/1079/1/012007

system). The remaining photoresist was stripped in a TePla 300-E plasma system (Technics Plasma, USA). Then, we processed the rear wafer side. First, the desired thickness of the membrane material (Fig.1-g), either Al or a tri-layer composed by Ti-Ni-Au, was sputtered using the Kenosistec KSH800 or the Material Research Corporation 903 equipment, employed respectively for aluminium and gold based metallization. Nickel and titanium, or a chromium layer, is standardly used to improve the intrinsic poor attachment of gold over SiO_2. Following back-side metallization, a 2 μm thick layer of photoresist AZ 6512 was spin-coated and exposed to UV light (not represented) using a Canon PLA600 mask aligner through a photomask allowing for patterning of the rear side of targets. The pattern is finally shaped by either wet etching of Al or of Ti-Ni-Au. The last step of the whole process is bulk Si micromachining (Fig.1-h) on the front wafer side necessary to open windows for the laser-interaction with the membranes. This is achieved by anisotropic etching in 40% KOH at 80°C during about 10 hours. The process relies on the fact that KOH has different etching rates on silicon crystal planes: it is fastest for (100) and (110) planes and lowest for (111) resulting in the characteristic "V" shaped 3D structures like truncated pyramids. Si micromachining is spontaneously stopped by the underlying buried oxide. To protect the metalized rear wafer side during Si micromachining a uniform layer of wax was deposited and baked during 1 h at 100°C. The wax dissolves easily in toluene and can be removed just before target delivery. Finally, for some samples we etched the underlying SiO_2 (Fig.1-i) by dip HF. Thin film thickness measurements were realized by using the Nanospec AFT 200 by Nanometrics for at least 5 different locations of each processed wafer; the final result is the average of these points. Optical microscopy characterization of membranes was accomplished by using the optical microscope by LEICA model DMLM. Analysis of surface roughness has been done with the atomic force microscopy by Veeco IV dimension, and scanning for each sample an area of 1 μm×1 μm. Aluminium membranes were fabricated having three thicknesses: 0.25 μm, 0.65 μm and 1 μm whereas in the case of gold based ones, we considered only two, 70 nm and 100 nm, which were both adhered over 20 nm titanium and 30 nm of nickel. For some cases, membranes were used tethered on the SiO_2, or the oxide layer was removed by dip HF. Each wafer located a total of 640 thin-layer membranes, out of which 384 have a surface area of 1 mm^2, and the remaining, of 0.5 mm^2. It is clear that the smaller the surface area, the greater will be the number of available targets for each wafer and consequently the lower will be the cost of each consumable membrane. Also, the robustness of submicrometric material layers is higher for reduced surfaces. However, this should be considered alongside the requirements for laser alignment and focusing, a part that for damage propagation at higher laser intensity than considered here. For this reason, we chose to fabricate two different area membranes and to analyse the implications on laser focusing which will be discussed in the following section. In general, it is important to remark that this is not ideal from the fabrication point of view because the etching time differs according to the structure size. For example, in our case we need to apply a supplementary etching time in order to complete Si micromachining of 0.25 mm^2 membranes, while keeping the others protected from the extra etching time which, at already opened windows, would result in breakage of metallic membranes. For each wafer, the fabrication yield was calculated as the difference between the numbers of intact membranes out of all target cells (Fig.2 (a) and (b)), resulting in an average of about 0.6 for all the productions (30 wafers). Normally, target fabrication is achieved in a place different from the laser installation and safe target transport should be also accounted as a further step of target production. We observed that keeping the wax layer on the metallized side and storing the wafers between the protective plastics which are thermally adhered for wafer dicing allow for a comfortable and safe target transport. Release of individual target cells could be finally realized at the laser installation. In this way, membranes fabricated according to the strategy presented are delivered in a format which is easy to manipulate inside the interaction chamber and adaptable to different experimental setups and needs, making it possible moreover to try several experimental conditions (thickness, material composition, local surface treatment) at once.

IOP Conf. Series: Journal of Physics: Conf. Series **1079** (2018) 012007 doi:10.1088/1742-6596/1079/1/012007

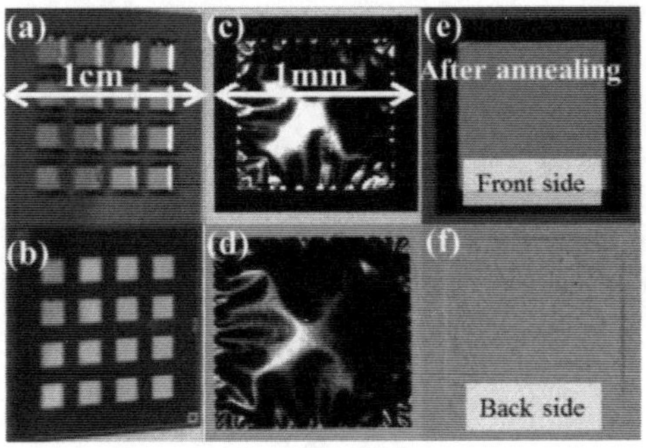

Figure 2. Back (a) and front (b) side of an individual target cell loading 16 membranes. Compressive stress shown by membranes following fabrication (c) and (d), and its release to tensile stress after annealing (e) and (f).

It is common to find that after fabrication, membranes exhibit compressive stress which is characterized by the presence of worm-like folding of the surface such as that in Fig. 2(c)-(d).The emergence of stress in membranes has been subject of much research in the field of CMOS technologies for the role they have in the correct functioning of many semiconductor based devices, and several methods for stress compensation have been proposed [5], [6]. In our case, surface folding could lead to difficulties for laser focusing and alignment, as well as for shot-to-shot reproducibility as we will discuss in the next section. For this reason, it is preferable to have membranes in a state of tensile stress causing material elongation so that their surfaces appear flat such that of Fig. 2(e)-(f). To achieve release of compressive stress into tensile one, we applied a short annealing at 400°C during 90 sec. which was found effective both for aluminium and gold based membranes. This could be done at the end of the fabrication right before using targets. We could observe that for free-standing membranes, i.e., that obtained after SiO_2 etching, compressive stress spontaneously releases after dip HF process. Thereafter, we characterized the membranes' surface roughness by using AFM which results are shown in Fig. 3. The picture shows the nanostructured character of the deposited film also referred as "cauliflower morphology" due to the fact that larger grains are formed by the aggregation of smaller ones. Grains size changes with film thickness.

Surface roughness is important for focus determination through the back reflection of visible laser light (speckle). In fact, in order to apply this method, as explained in [7], it is necessary that surface roughness be in the range of the laser wavelength. For all analysed samples we are far from reaching this value. For the thickest ones, which are expected to be also the roughest ones, we could only measure an average roughness of about 100 nm whereas for all remaining cases it stays in the range of a few nanometers. For this reason, in order to determine the focus position through the speckle we glued on top of each target cell a slice of aluminium foil allowing for clear visualization of the interference pattern. However, in order to account for the effects of inaccurate focus determination we mapped proton energy within a distance of ±100 μm from the focus estimated initially.

IOP Conf. Series: Journal of Physics: Conf. Series **1079** (2018) 012007 doi:10.1088/1742-6596/1079/1/012007

Figure 3. Roughness characterization of an aluminium membrane 1 µm thick through AFM.

3. Proton acceleration experiments

All experiments were conducted at a 3 TW/55 fs table-top laser system having 165 mJ pulse energy behind the compressor. It can operate both in single shot mode and up to 100 Hz repetition rate. The system is based on chirped pulsed amplification (CPA) and it has been developed specifically for laser-proton acceleration by one of the partners [4]. Following compression, laser pulses are injected into the interaction chamber kept at a pressure of about 10^{-2} mbar which is sufficient to prevent laser filamentation and deceleration of protons before reaching particle detectors. Laser pulses are then focused on the target by a *f/3* gold coated off-axis parabolic mirror (OAP) at an incidence angle of 30°. A 632 nm He:Ne laser is coupled into the same optical path of the principal Ti:Sa beam and used to control the target positioning in transverse direction. Characterization of accelerated particles was achieved by means of CR-39 plates placed at 100 cm distance behind the target, and a time-of-flight (TOF) detector based on a fast plastic scintillator located at a distance of 237 cm from the target. Both detectors were calibrated with standard particle beams at the National Accelerator Centre (CNA) at Seville. Proton acceleration experiments were performed in single shot modality. Individual target alignment and focusing has been done for the first membrane of each target cell used. All data were taken with 1 mm^2 membranes. For a series of shots within the same target cell we assumed that the focus does not vary from one membrane to the subsequent ones and thus maintained a reference position. In order to study the dependency of the proton energies on the longitudinal position we moved away from the established focus in steps of ±50 µm. Thereafter, we obtained the maximum proton energy for aluminum and gold bases membranes of different thicknesses for both free-standing and SiO_2 tethered membranes. For each thickness, the maximum proton energy is calculated averaging over the highest values of energy measured in the focus position. The values obtained are coherent with previous results showing that with high contrast ($>10^8$) thinner targets are better energy converters [8]–[13]. An example of obtained results is shown in Fig. 4(a) for gold based membranes. This is the typical energy distribution around the focus position obtained for each analyzed thickness, independently from the material used. Although for many specimens a parabolic shape around the zero position has been observed, the corresponding energy variations within ±50 µm account for only 10% of the value measured in the exact focus. This is interesting in view of the development of applications requiring high rate repetition operability. In fact, whereas at low frequency it is possible to perform even single target alignment, at high frequency this possibility is lost so that it is important that off-focus energy values suffer only slight variations. For both aluminum and gold based membranes, we could observe that data from free-standing and SiO_2 tethered membranes differ by 5-10% (Fig. 4(b)).

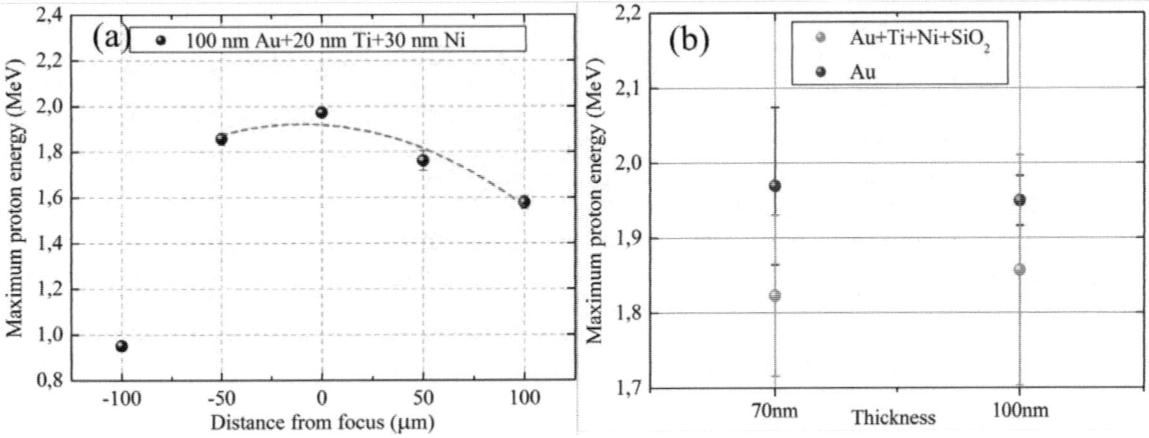

Figure 4. (a) Maximum proton energy distribution around focus position. (b) Comparison between maximum proton energy obtained with free-standing and SiO_2 tethered membranes.

The higher energy with pure Au may be intuitive because the membrane with oxide layer is significantly thicker. In addition, in our set-up, the laser pulse interacts first with the metallic, and then with the semiconductive layer which could prevent the effective propagation of the electric sheath inside the target. However, these findings deserve a deeper analysis with less parameter variability and a better contrast over ASE. The only observation we can affirm at present is that under current working conditions, the presence of the SiO_2 behind the metallic membrane is not especially relevant, while avoiding the dip HF etching of this layer can in general improve the yield of fabrication.

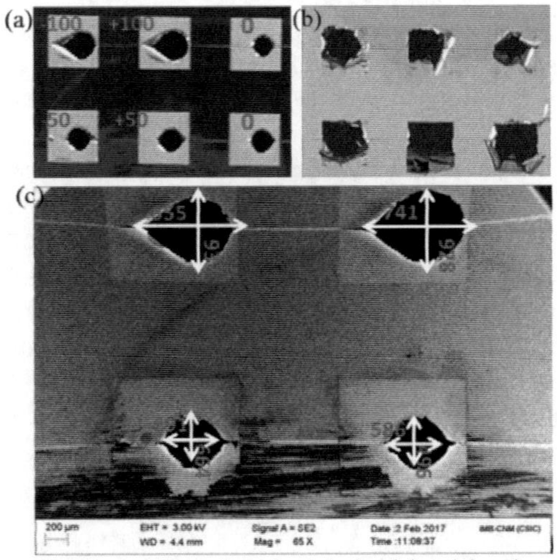

Figure 5. Comparison between shock propagation in (a) double side photolithography membranes and (b) all-metalized rear side membranes. (c) Astigmatism in tracks left by laser shots off-focus.

Another interesting remark arose from the post-shot target analysis between membranes belonging to target cells with photolithography also on the back wafer side (see Fig. 5 (a)) and bottom (Fig. 5 (b)).Although we could observe, as represented in Fig. 5 (b), that membranes after shots look more damaged when they belong to a double-side photolithography target cell. This phenomenon should be investigated at higher laser intensity, because it is well-known that target damage and "fratricide" are not an issue below 1 J. It is also interesting to observe that being completely destroyed after interaction, these membranes do not show tracks of the off-focus alignment chosen in some case to study proton energy dependence, which are on the other side appreciable on membranes of Fig. 5 (a).

The progressive off-focus introduced alters the shape of the hole left by laser which becomes more elliptical and tends to have a bigger area than in the case of membranes shot in the focus position. In general, laser damage propagates over an area which is two orders of magnitude bigger (Fig. 5 (c)) than the laser-spot area estimated to be $\approx 5.0\times11.5\ \mu m^2$ (FWHM). The size of damage propagation is significant with respect to mask design because it limits from below the opening area available for laser interaction. This has also consequences for the array density one can consider which becomes relevant for high repetition rate operability.
Finally, we compared data obtained from MEMS based membranes with data recorded from plain foils of the same thickness and material (aluminum) from benchmark experiments conducted with the same laser system [4]. According to our results, the use of MEMS membranes improves the proton acceleration performance of a 30% (in case of 0.65 μm thick targets) and of a 9.8% (in case of 1 μm thick targets) [14]. Here too, further studies with higher statistics will be necessary to corroborate these findings.

4. Conclusions

We have presented the fabrication of submicrometric, conductive, thin-layer membranes embedded in a Si wafer according to a MEMS based manufacturing approach. This combines standard photolithography, material layer deposition, selective etching and bulk Si micromachining. Photolithography is used to rapidly pattern the Si surface in such a way to define windows on the front wafer side having the same area of metallic membranes deposited on the rear wafer side. Finally, Si micromachining allows for opening windows on the front side thus making membranes, either free-standing or tethered on a nanometric SiO_2 layer, accessible for laser interaction. Gold, and aluminum, based membranes were fabricated with variable thicknesses below 1 μm and were used as targets for laser-driven proton acceleration through the TNSA mechanism showing better efficiency in energy conversion compared to standard foil targets used for benchmark experiments at the same laser installation.
MEMS based manufacturing offers many advantages for target fabrication as compared to different methods which mainly rely on manual and individual target arrangements. Among these, the reproducibility of the fabrication thanks to parallel processing, the controllability of membrane thickness and composition over a wide range of options, and the possibility of target manufacturing over large area substrates by using conventional microfabrication facilities. Indeed, the degree of target complexity, for example, through the introduction of 3D surface modification could be straightforwardly considered. All these elements would foster the cheap fabrication of simple and even complex targets useful for high-power, high repetition rate lasers experiments and would contribute to the development of applications based on ultra-intense laser-plasma interactions.

Acknowledgements

This work has made use of the Spanish ICTS Network MICRONANOFABS partially supported by MEINCOM. This project has been financed by the Spanish Ministry for Economy and Competitiveness within the Retos- Colaboración 2015 initiative, ref. RTC-2015-3278-1. P. Mur has received a grant of the Garantía Juvenil 2015 program.

References

[1] I. Prencipe, J. Fuchs, S. Pascarelli, D. W. Schumacher, R. B. Stephens, N. B. Alexander, R. Briggs, M. Büscher, M. O. Cernaianu, A. Choukourov, M. De Marco, A. Erbe, J. Fassbender, G. Fiquet, P. Fitzsimmons, C. Gheorghiu, J. Hund, L. G. Huang, M. Harmand, N. J. Hartley, A.

IOP Conf. Series: Journal of Physics: Conf. Series **1079** (2018) 012007 doi:10.1088/1742-6596/1079/1/012007

Irman, T. Kluge, Z. Konopkova, S. Kraft, D. Kraus, V. Leca, D. Margarone, J. Metzkes, K. Nagai, W. Nazarov, P. Lutoslawski, D. Papp, M. Passoni, A. Pelka, J. P. Perin, J. Schulz, M. Smid, C. Spindloe, S. Steinke, R. Torchio, C. Vass, T. Wiste, R. Zaffino, K. Zeil, T. Tschentscher, U. Schramm, and T. E. Cowan, "Targets for high repetition rate laser facilities: needs, challenges and perspectives," *High Power Laser Sci. Eng.* 5, e17, 2017.

[2] M. Roth and M. Schollmeier, "Ion Acceleration - Target Normal Sheath Acceleration," Proceedings of the CAS-CERN Accelerator School: PlasmaWake Acceleration, Geneva, Switzerland, http://dx.doi.org/10.5170/CERN-2016-001.231,2017.

[3] A. Macchi, M. Borghesi, and M. Passoni, "Ion acceleration by superintense laser-plasma interaction," *Rev. Mod. Phys.*, 85(2), 751 (43), 2013.

[4] P. Bellido, R. Lera, M. Seimetz, A. Ruiz-De La Cruz, S. Torres-Peirò, M. Galán, P. Mur, I. Sánchez, R. Zaffino, L. Vidal, A. Soriano, S. Sánchez, F. Sánchez, M. J. Rodríguez-Álvarez, J. P. Rigla, L. Moliner, A. Iborra, L. Hernández, D. Grau-Ruiz, A. J. González, J. J. García-Garrigos, E. Díaz-Caballero, P. Conde, A. Aguilar, L. Roso, and J. M. Benlloch, "Characterization of protons accelerated from a 3 TW table-top laser system .," *JINST*, 12 p. T05001, 2017.

[5] R. F. W. A M Ghaderi, N P Ayerden, G de Graaf, "Minimizing stress in large-area surface micromachined perforated membranes with slits," *J. Micromechanics Microengineering*, vol. 25, p. 074010(9 p), 2015.

[6] D. Jaeggi and B. Henry, "Thermal Converters by CMOS technology," PhD thesis, ETH Zurich, 1996.

[7] F. Lindau, *Laser-driven particle acceleration*. Ph.D. thesis, Lund University, 2007.

[8] S. Abedi, D. Dorranian, M. E. Abari, and B. Shokri, "Relativistic effects in the interaction of high intensity ultra-short laser pulse with collisional underdense plasma," *Phys. Plasmas*, vol. 18, no. 9, pp. 1–6, 2011.

[9] M. Kaluza, J. Schreiber, M. I. K. Santala, G. D. Tsakiris, K. Eidmann, J. Meyer-Ter-Vehn, and K. J. Witte, "Influence of the laser prepulse on proton acceleration in thin-foil experiments," *Phys. Rev. Lett.*, 93 (4), 045003, 2004.

[10] P. Antici, J. Fuchs, E. D'Humières, E. Lefebvre, M. Borghesi, E. Brambrink, C. A. Cecchetti, S. Gaillard, L. Romagnani, Y. Sentoku, T. Toncian, O. Willi, P. Audebert, and H. Pépin, "Energetic protons generated by ultrahigh contrast laser pulses interacting with ultrathin targets," *Phys. Plasmas*, vol. 14, no. 3, 2007.

[11] T. Ceccotti, A. Lévy, H. Popescu, F. Réau, P. D'Oliveira, P. Monot, J. P. Geindre, E. Lefebvre, and P. Martin, "Proton acceleration with high-intensity ultrahigh-contrast laser pulses," *Phys. Rev. Lett.*, 99, 185002, 2007.

[12] A. P. L. Robinson, P. Foster, D. Adams, D. C. Carroll, B. Dromey, S. Hawkes, S. Kar, Y. T. Li, K. Markey, P. McKenna, C. Spindloe, M. Streeter, C. G. Wahlström, M. H. Xu, M. Zepf, and D. Neely, "Spectral modification of laser-accelerated proton beams by self-generated magnetic fields," *New J. Phys.*, vol. 11, 2009.

[13] M. Nishiuchi, H. Daido, A. Yogo, S. Orimo, K. Ogura, J. Ma, A. Sagisaka, M. Mori, A. S. Pirozhkov, H. Kiriyama, S. V. Bulanov, T. Z. Esirkepov, I. W. Choi, C. M. Kim, T. M. Jeong, T. J. Yu, J. H. Sung, S. K. Lee, N. Hafz, K. H. Pae, Y. C. Noh, D. K. Ko, J. Lee, Y. Oishi, K. Nemoto, H. Nagatomo, K. Nagai, and H. Azuma, "Efficient production of a collimated MeV proton beam from a polyimide target driven by an intense femtosecond laser pulse," *Phys. Plasmas*, vol. 15, no. 5, 2008.

[14] R. Zaffino, M. Seimetz, I. Sanchez, P. Mur, D. Quirion, P. Bellido, R. Lera, L. Martin, M. Lozano, and G. Pellegrini, "Efficient proton acceleration from a 3 TW table-top laser interacting with submicrometric mass-produced solid," *under Review*.

IOP Conf. Series: Journal of Physics: Conf. Series **1079** (2018) 012008 doi:10.1088/1742-6596/1079/1/012008

Optimization of a fast rotating target to produce kHz X-ray pulses from laser-plasma interaction

L Martín[1], J Benlliure[1], D Cortina[1], J J Llerena[1], D González[1] and C Ruiz[2]

[1] IGFAE, Universidade de Santiago de Compostela, 15782 Santiago de Compostela, Spain.
[2] Instituto Universitario de Física Fundamental y Matemáticas y Departamento de Didáctica de la Matemática y de las Ciencias Experimentales, Universidad de Salamanca, Patio de Escuelas s/n, Salamanca, Spain.

E-mail: lucia.martin@usc.es

Abstract. We report the development of a fast rotating target to produce ultrashort incoherent X-ray pulses from bremsstrahlung. These short X-ray pulses are produced in the laser-plasma interaction of a 35 fs, 1 mJ pulse of a Ti:Sa laser of 1 kHz repetition rate, with a solid metallic target. In this paper, we report our developments to improve the stability of this micron size source of ultrashort X-rays. As the Rayleigh length is very small (<15 μm), wobbling of the rotatory stage can reduce the intensity on target and change the characteristics of the source. We describe the methods we have developed to measure and adjust the stability of the focus on target. These advances are important for the development of sources with high average power and good stability. The X-ray source has a broad Maxwellian-like distribution with temperatures of around 10 - 40 KeV and could be used for advanced X-ray imaging such as absorption or phase contrast tomography.

1. Introduction

Femtosecond hard X-ray sources are an important tool to describe the ultrafast dynamics in physics [1, 2], chemistry [3], biology [4] and material science [5]. Optical pump and X-ray probe have demonstrated to be an important test for the sensitivity of detection schemes and provide structural and dynamical information of the systems under study.
Laser-driven table-top X-ray sources have been demonstrated in solid [6, 7, 8] and gas media [4, 9], and a new family of all optical sources are being developed with high power lasers [10]. Also, compact X-ray sources based on the λ^3 regime [11, 12] have been demonstrated in a wide variety of solid targets. These sources operate with a modest energy of few mJ and can reach KHz repetition rates which increase the average power of the source [13].

These X-ray sources are based on the interaction of an ultrashort laser pulse with an over-dense plasma. Combination of laser field and ionization during the first cycles of the laser leads to the creation of a plasma on the target surface [14]. The density of this plasma is several times the critical density, containing hot electrons with relativistic velocities due to the high intensity of the pulse ($I > 10^{16}$ W/cm^2). These electrons interact with the heavy ions of the target as they leave it, producing Bremsstrahlung and characteristic radiation [15, 16, 17]. These collisions lead to the heating of the target, increasing the plasma pressure and producing the ablation of the target surface [14].

IOP Conf. Series: Journal of Physics: Conf. Series **1079** (2018) 012008 doi:10.1088/1742-6596/1079/1/012008

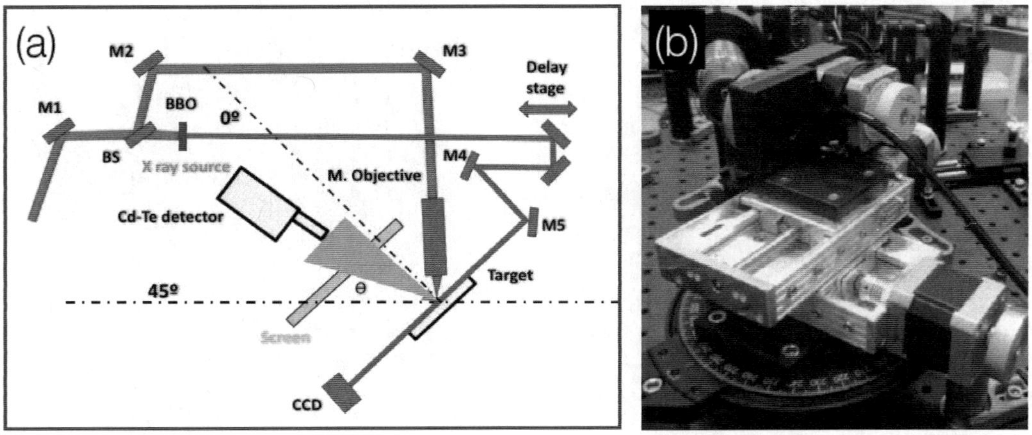

Figure 1: (a) Diagram of the experimental setup. (b) Picture of the linear and rotating stages of the target system.

As the laser drives the electrons only for a short period of time, the result is a well defined beam of X-ray pulses with a broad energy distribution and a short duration of hundreds of femtoseconds [18].
In this paper, we focus on the development of a fast rotating target which can sustain a stable and robust X-ray source in the λ^3 regime [19, 20]. The final objective is to increase the average power of the source and to obtain long operation periods supporting real world applications that would benefit special characteristics of this type of X-ray source [21]. The mechanical stability of the target and the implementation techniques to diagnose the position of the target with respect to the laser focus, are the key issues to produce a X-ray source with these characteristics.

2. Experimental set up and target positioning system for the kHz X-ray source

A back-scattered X-ray laser driven source has been recently installed at the Laser Laboratory for Acceleration and Applications (L2A2) at the University of Santiago of Compostela. We report the production of X-ray pulses at 1 KHz repetition rate from the relativistic interaction of an intense laser pulse and a solid target in air. The target is a flat copper metallic plate (55 mm x 55 mm) of 1 mm thickness with a mirror like surface where the laser pulse impacts. In order to have a stable X-ray source, the target needs to accomplish two conditions simultaneously: The positioning system should refresh the target material, because the material is destroyed after each shot. During the interaction, a large number of electrons are expelled from the material leading to the destruction of the target of an area much larger than the laser focus. Therefore, the target positioning system should rapidly replace the material to sustain a KHz repetition rate. The positioning system should also keep the target at the focal plane of the laser as the material is refreshed. Because we use a tight focusing scheme to obtain the smallest possible focal spot, the Rayleigh length is small ($<15\ \mu m$) and the tolerance in positioning must be kept below the Rayleigh length.

The experimental setup is shown in Fig.2. An ultrashort laser pulse (THALES ALPHA 10/XS, 1 mJ, 35fs, 1KHz, 800 nm) is first divided through a beam-splitter (BS, 90R:10T) and then transported with M1, M2 and M3 dielectric mirrors (Thorlabs, BB1-E03) towards the microscope objective (Mitutoyo, M Plan Apo NIR 20X) with a focal length of 10 mm and Rayleigh length $<15\ \mu m$.

IOP Conf. Series: Journal of Physics: Conf. Series **1079** (2018) 012008 doi:10.1088/1742-6596/1079/1/012008

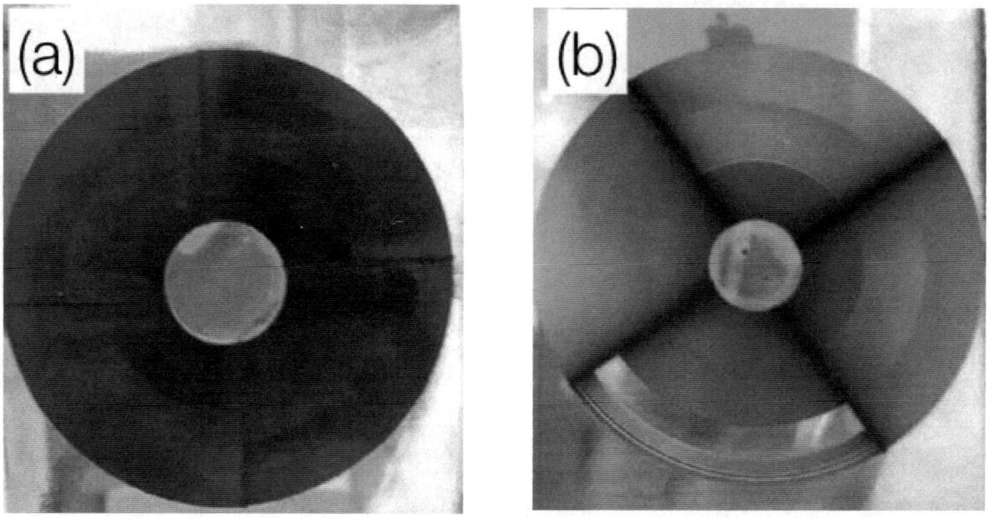

Figure 2: Two targets irradiated in different conditions. (a) Target irradiated with constant angular velocity and (b) variable angular velocity.

The laser pulse is focused on target at 45° to produce a back-scattered X-ray beam close to the normal direction to the target. The diameter of the focal spot is estimated to be 3.5 μm, but a 5 μm defocussing results in bigger spots up to 3.8 μm. In addition to this, the incidence of the laser beam is at 45° so the effective spot diameter is between 4.9 and 5.4 μm. These spot sizes result in intensities on target between 1.1x10^{17} W/cm^2 and 9.4x10^{16} W/cm^2. The X-ray beam is analyzed with a Cd-Te spectrometer (Amptek, XR-100T-CdTe) to measure the X-ray energy spectrum. We use a 0.3 mm aluminum attenuator and a 1.15 mm lead collimator to minimize pile-up effects and to avoid the laser beam to entering detector.

The target positioning system and the diagnostic of the position are two of the most critical elements of the setup. The performance and stability of the X-ray source depends on the material refreshing and positioning of the target on focus during source operation. There are several solutions that have been used for this problem [2, 3, 8, 10, 9]. We have chosen a positioning and refreshment system that consists on two linear stages (PiMicos VT-80) to change the longitudinal and perpendicular focal position on target, and one rotation stage (PiMicos DT-50). The target is attached to the rotating stage which is perpendicular to the optical board. The rotating stage is also attached through an L-shape piece to the two linear stages as shown in the Fig.1. For the measurements presented in this paper, we use laser pulses with an energy of $E = 800\ \mu J$ focused on target at 45° incidence. The target material is fixed to the rotatory stage with double side tape. In this way, we minimize wobbling in the external radii of the pattern made by the movement of the target.

As the laser impacts on target it damages a region whose size depends on the energy and angle of incidence of the laser, being in any case larger than the laser spot size on target. As the rotation stage moves, the impacts on the target draw a circle as the stage complete a turn. Then, the first linear stage moves the target perpendicularly to the focal plane, in such a way that the next round of impacts draws a new concentric circle.

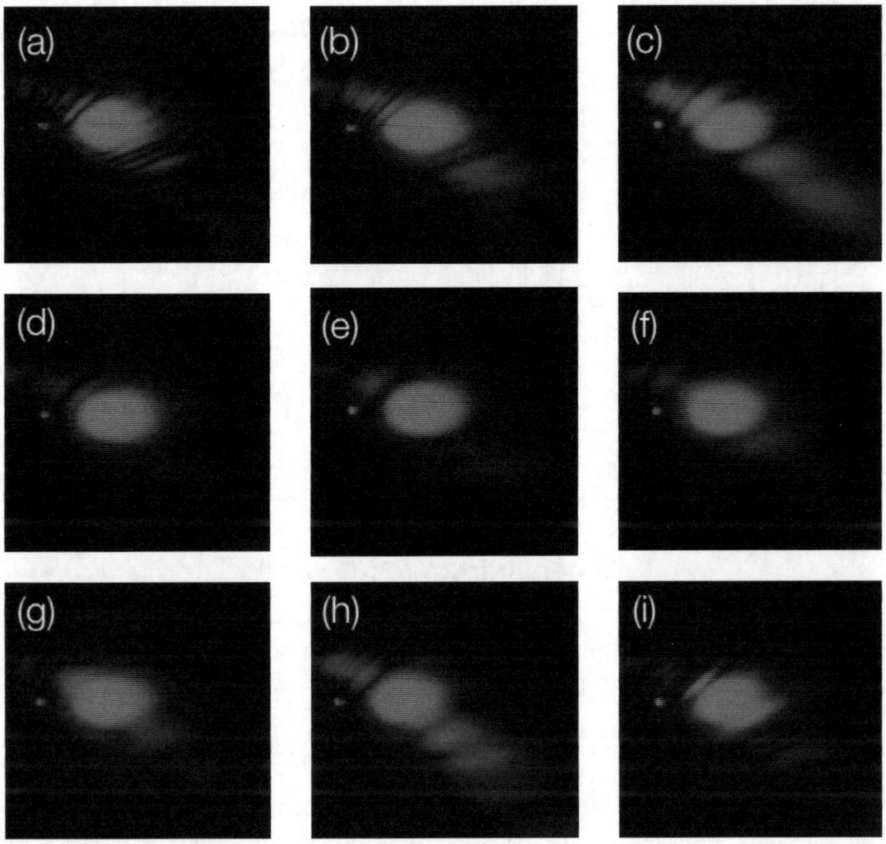

Figure 3: Speckle at different target positions. (a) 100 μm after focus, (b) 50 μm after focus, (c) 20 μm after focus, (d) 5 μm after focus, (e) Speckle at focus, (f) 5 μm before focus, (g) 20 μm before focus, (h) 50 μm before focus and (i) 100μm before focus.

The second linear stage moves the target in and out of focus and it is used only to place the target in focus. This rotating target solution has important advantages compared to a Cartesian arrangement which is frequently used. In the case of a rotating target, there are no big accelerations in the movement, which favor the stability of the source. Moreover, linear stages are not as fast as rotating ones, so the setup presented here allows to minimize superposition of laser impacts at the target surface at 1kHz repetition rate, optimizing the X-ray dose. The total number of shots per target depends on the distance between impacts and the available area on target. To maximize the number of shots per target and optimize the performance of the X-ray source we have established two strategies: The first strategy keeps the angular velocity of the target constant (ω=15°/s), which means that the linear velocity depends on the radius of the impact from the rotation center ($v = r\omega$). With this strategy the distance between impacts decrease with the radius. In the second strategy, we keep the linear velocity constant ($v = 196$ mm/s). As the perpendicular motor moves the target towards smaller radius, we change the angular velocity $\omega(r) = v/r$ using velocities between $450°s^{-1}$ and $900°s^{-1}$ to keep the separation between impacts constant.

IOP Conf. Series: Journal of Physics: Conf. Series **1079** (2018) 012008 doi:10.1088/1742-6596/1079/1/012008

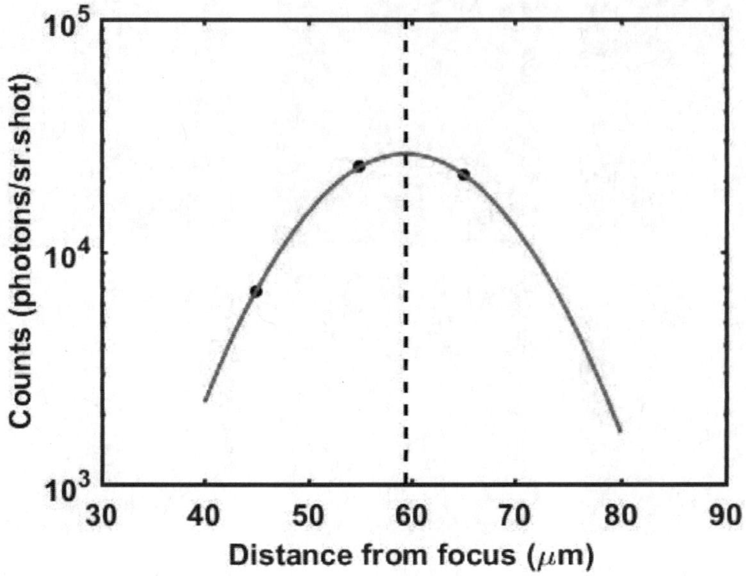

Figure 4: Displacement of the nominal focal position due to self-focusing. The total counts for three spectra measured at different target positions are shown and used to calculate the best position (59 μm).

Figure 2 shows two copper plates after being used as targets. Fig.2(a) corresponds to an irradiation following the first strategy of constant angular velocity, and Fig.2(b) shows the case of constant linear velocity. In both cases, the uniformity in colors shows that the source is always in focus producing a similar damage per shot. Irradiation at constant linear velocity requires the measurement of the initial radius where the laser impacts. We use two laser impacts on target separated by a 180° rotation to determine the center of the rotatory stage. Operating with a constant angular velocity is simpler, because it is not necessary to know the radius. However, the X-ray production in this mode changes from one radius to another, because the linear velocity varies, and also the superposition of laser impacts on target surface increases as the radius decreases. In order to reach a stable operation in this mode, it would be necessary to use larger targets to avoid overlap between impacts. We discuss the effect on the spectra of the two operation modes in Section 4.

3. Diagnostics for target positioning

A stable X-ray source for this high repetition rate requires the minimization of the relative motion of the target surface with respect to the laser focus as the target rotates. The Rayleigh length is small ($<$15 μm), so we need to position the target with an accuracy better than 5 μm to have the same intensity on target for each shot.

We use a speckle technique, to optimize the relative position of the target surface and the laser focus. For this purpose, we use a visible He-Ne laser (633nm, 1.2 mW, continuum) following the same path as the main infrared beam, and focused on target by the microscope objective. The specular reflection of the visible laser (speckle) is an homogeneous spot if the target is on focus (Fig.3(e)), because a 5 μm diameter spot is smaller than the roughness of target surface.

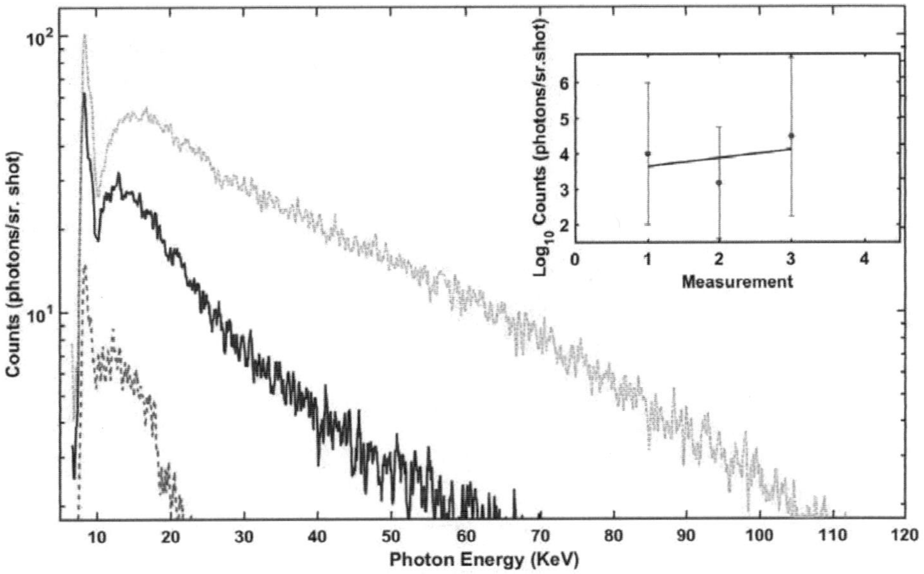

Figure 5: Three energy spectra for different radii and I_{laser}=1.2x10^{17} W/cm^2,Cd-Te distance 24.5 cm and angle 15° in constant angular velocity operation mode. (Inset image) Variation of spectra total counts 55%.

However, if the target is not on focus, the laser spot on the target surface will be larger and we will see the structure of the surface on to the speckle, becoming inhomogeneous (Figure 3). To position the target in focus, we observe the image of the He-Ne speckle on a white screen with a CMOS camera (5 Mpx, Mightex) while moving the target position with the longitudinal motor. The best focal position is determined when the cleanest speckle is found.

However, the implementation of this technique in air is more complicated, because the compressed infrared laser pulse (leaving the objective) suffers a nonlinear interaction with air before reaching the target surface. This nonlinear interaction makes the infrared laser beam converge before the visible one, resulting in the self-focusing of the infrared pulse due to the Kerr effect. In order to determine the best focal position, several spectra are taken at different distances from the He-Ne best speckle position.
We define the best focus position as the distance that maximises the total number of photons in the spectrum, while taking into account the self-focusing. This effect depends on the intensity, so for intensities around 10^{17} W/cm^2, the required displacement from the speckle position is about 59 μm towards the objective as shown in Figure.4.

4. X-ray source performance

As a result of the laser-plasma interaction, a well defined X-ray conic beam is emitted from the target in a direction close to the target normal. The technique described in Sect.3 for the target-focus positioning has been used to investigate source stability at different radial positions. The X-ray spectrum is measured with a Cd-Te spectrometer at 15°-17° from the target normal.

IOP Conf. Series: Journal of Physics: Conf. Series **1079** (2018) 012008 doi:10.1088/1742-6596/1079/1/012008

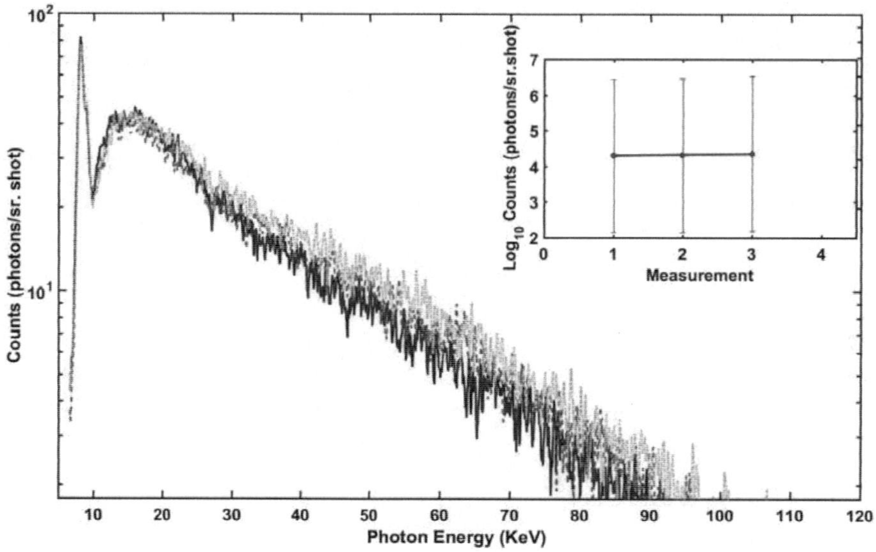

Figure 6: Three energy spectra showing the stability of the source taken at different radii and I_{laser}=1.1x10^{17} W/cm^2,Cd-Te distance 24.5 cm and angle 17°) and variable angular velocity operation mode. (Inset image) Variation of spectra total counts.

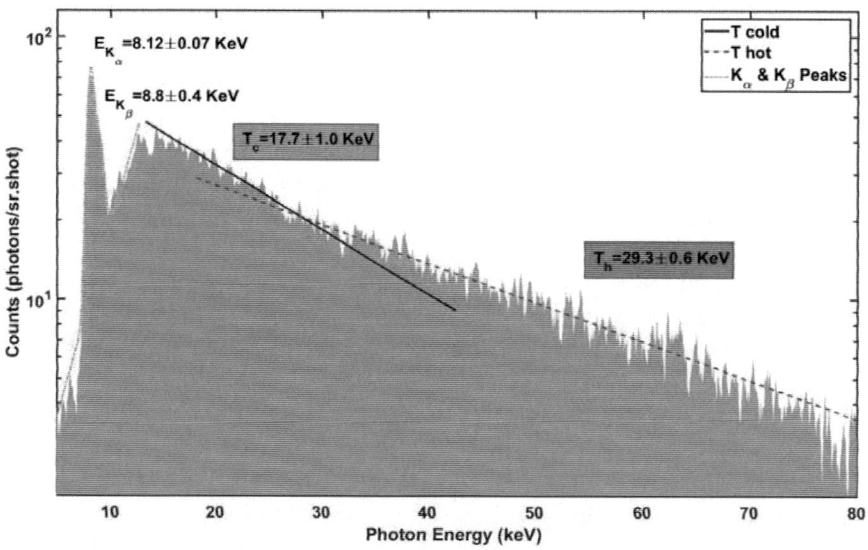

Figure 7: Typical energy spectrum obtained irradiating a Cu target. In the spectrum we can clearly observe the Cu K peaks and also the cold and hot temperatures of the X-Ray continuum.

IOP Conf. Series: Journal of Physics: Conf. Series **1079** (2018) 012008 doi:10.1088/1742-6596/1079/1/012008

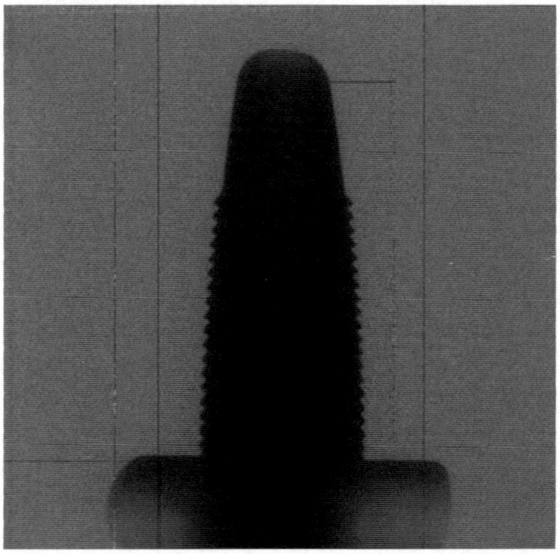

Figure 8: X-Ray Absorption Image taken with the X-ray source. The image is obtained after irradiating 30 seconds at 1 Khz.

We can see the effect on spectra obtained with the same conditions and constant angular velocity in Figure 5, where deviation in the total number of between the measurements at different radii is above 50%. However, if we adjust the rotation speed to the radial position of the laser impact on target, we obtain a very stable X-ray source. Indeed, Figure 6 illustrates that the reproducibility of the X-ray energy spectra in this case is better than 6%.
In Figure 7, we report a typical X-ray energy spectrum, obtained with a Cd-Te spectrometer at 17° from target and a laser intensity of $1.1 \mathrm{x} 10^{17}$ W/cm^2. The observed spectrum presents two peaks of copper characteristic radiation, K_α and K_β and a Bremsstrahlung continuum with a bi-Maxwellian like distribution. This distribution consists on two temperatures T_{cold} and T_{hot} which are related with the different dynamics of the electrons inside the over-dense plasma. We obtain 8.12 and 8.8 keV for Cu K peaks, and T_{cold}=17.7 keV and T_{hot}=29.3 keV.

5. Foreseen applications

These kind of novel X-Ray sources driven by laser-plasma interaction open new possibilities of high quality X-Rays at small facilities. The principal advantages of laser-plasma X-ray sources are the smaller size of the source, and the short duration of the X-ray pulses. The short duration of the X-ray pulses (hundreds of femtoseconds to picoseconds) makes possible to perform X-ray time resolved images, which is interesting in a large number of research fields. Compared with conventional X-Ray tubes used for imaging, the smaller source size increases the spatial coherence of the emitted X-rays, making possible to implement more sophisticated imaging techniques than absorption imaging such as phase contrast imaging. Indeed, the quality of conventional absorption images is also improved, because the increased spatial coherence decrease the image blurring, resulting in better image quality. In Figure 8, we can observe an absorption image performed with the presented X-ray source and a X-ray CCD array detector (RadEye 2, low dose:10-50 keV). Besides, the source stability achieved is good enough to perform also tomographic experiments. Moreover, the X-ray source is a pulsed source with very short duration, so it is possible to perform pump-probe experiments with infrared and X-rays simultaneously. It is clear that the field of application of these novel sources is growing up.

6. Conclusions

In this work, we have presented the new X-ray laser driven source recently developed at the L2A2. The source is based on a back-scattered X-ray emission of a hot over dense plasma induced by 1 mJ, 35 fs laser pulse on a Cu solid target. To guarantee material refreshment at 1 kHz repetition rate, the target is mounted on a fast rotatory positioning system. The target has to be placed on the laser focus during the movement, and in order to avoid the wobbling we have presented solutions for target mounting. We have implemented a speckle technique mixed with a best focus determination based on X-ray spectra, which allows us to take into account the effect of the laser self-focusing in air. We have also demonstrated a way to increase the source stability, operating with a variable rotation speed. A complete characterization of a X-ray spectrum is also presented, showing the characteristic Cu K peaks and the Bremsstrahlung continuum composed by two different temperatures (T_{cold}=17.7 keV and T_{hot}=29.3 keV) as it is characteristic in this kind of laser-plasma interaction. To finish, we have also shown the first application of our X-ray source with a X-ray absorption image. In future, we plan to obtain tomography and phase contrast images with the X-ray source presented here. However, the source presented here is not only interesting for imaging purposes, also for going deeper in the physics of the plasma, the different distributions of electrons and the interactions among the laser and the target.

Acknowledgments

This work has been supported by Spanish Ministry of Industry within the framework of the RETOS project (RTC-2015-3278-1) and by Xunta de Galicia and EU, LASERPET project (2013-AD009.01). Camilo Ruiz acknowledges the MINECO project FIS2016-75652-P.

References

[1] Wachulak P W, Bartnik A, Wegrzynski L, Kostecki J, Jarocki R, Fok T, Szczurek M and Fiedorowicz H 2013 *Nuclear Instruments and Methods in Physics Research* **B 311** 42–46
[2] Bonvalet A, Darmon A, Lambry J C, Martin J L and Audebert P 2006 *Opt. Lett.* **31** 2753–55
[3] Legall H, Blobel G, Stiel H, Sandner W, Seim C, Takman P, Martz D H, Selin M, Vogt U, Hertz H M, Esser D, Sipma H, Luttmann J, Hoefer M, Hoffmann H D, Yulin S, Feigl T, Rehbein S, Guttmann P, Schneider G, Wiesemann U, Wirtz M and Diete W 2012 *Opt. Express* **20** 18362–69
[4] Wenz J, Schleede S, Khrennikov K, Bech M, Thibault P, Heigoldt M, Pfeiffer F and Karsch S 2015 *Nat. Commun.* **6**
[5] Silies M, Witte H, Linden S, Kutzner J, Uschmann I, Förster E and Zacharias H 2009 *Applied Physics A* **96** 59–67
[6] Gobet F, Hannachi F, Aléonard M M, Chemin J F, Claverie G, Gerbaux M, Malka G, Scheurer J N and Tarisien M 2006 *Rev. Sci. Instrum.* **77**
[7] Hou B, Easter J, Mordovanakis A, Krushelnick K and Nees J A 2008 *Opt. Express* **16** 17695–705
[8] Li M, Huang K, Chen L, Yan W, Tao M, Zhao J, Ma Y, Li Y and Zhang J 2016 *Radiat. Phys. Chem.*
[9] Zhang L, Chen L M, Yuan D W, Yan W C, Wang Z H, Liu C, Shen Z W, Faenov A, Pikuz T, Skobelev I, Gasilov V, Boldarev A, Mao J Y, Li Y T, Dong Q L, Lu X, Ma J L, Wang W M, Sheng Z M and Zhang J 2011 *Opt. Express* **19** 25812–22
[10] Zamponi F, Ansari Z, Korff Schmising C v, Rothhardt P, Zhavoronkov N, Woerner M, Elsaesser T, Bargheer M, Trobitzsch-Ryll T and Haschke M 2009 *Appl. Phys.* **A 96** 51–58
[11] Naumova N, Sokolov I, Nees J, Maksimchuk A, Yanovsky V and Mourou G 2004 *Phys. Rev. Lett.* **93** 195003
[12] Naumova N M, Nees J A, Sokolov I V, Hou B and Mourou G A 2004 *Phys. Rev. Lett.* **92** 063902
[13] Weisshaupt J, Juvé V, Holtz M, Ku S, Woerner M, Elsaesser T, Ališauskas S, Pugžlys A and Baltuška A 2014 *Nat. Photonics*
[14] Gibbon P 2005 *Short Pulse Laser Interactions with Matter: An Introduction* (London: Imperial College Press) ISBN 1-86094-135-4 record converted from VDB: 12.11.2012 URL `http://juser.fz-juelich.de/record/47943`
[15] McCall G H 1982 *J. Phys.* **D**: *Appl. Phys* **15** URL `http://iopscience.iop.org/0022-3727/15/5/012`
[16] Salzmann D, Reich C, Uschmann I, Förster E and Gibbon P 2002 *Physical Review* ***E*** **65**
[17] Galy J, Maucec M, Hamilton D J, Edwards R and Magill J 2007 *New J. Phys.* **9**

IOP Conf. Series: Journal of Physics: Conf. Series **1079** (2018) 012008 doi:10.1088/1742-6596/1079/1/012008

[18] Kieffer J C, Chaker M, Matte J P, Pépin H, Côté C Y, Beaudoin Y, Johnston T W, Chien C Y, Coe S, Mourou G, and Peyrusse O 1993 *Physics of Fluids* **B**: *Plasma Physics* **5** 2676–81
[19] Hou B, Nees J, Mordovanakis A, Wilcox M, Mourou G, Chen L M, Kieffer J C, Chamberlain C C and Krol A 2006 *Appl. Phys.* **B 83**
[20] Hou B, Nees J A, Theobald W, Mourou G A, Chen L M, Kieffer J C, Krol A and Chamberlain C C 2004 *Appl. Phys. Lett.* **84** 2259–61
[21] Jiang Y, Lee T, Li W, Ketwaroo G and Rose-Petruck C G 2002 *Opt. Lett.* **27** 963–65

IOP Conf. Series: Journal of Physics: Conf. Series **1079** (2018) 012009 doi:10.1088/1742-6596/1079/1/012009

Development of a multi-shot experiment for proton acceleration

C Ruiz[1], J Benlliure[2], D Cortina[2], D González[2], J Llerena[2] and L. Martín[2].

[1] Instituto Universitario de Física Fundamental y Matemáticas y Departamento de Didáctica de la Matemática y de las Ciencias Experimentales, Universidad de Salamanca, Patio de Escuelas s/n, Salamanca, Spain.

[2] IGFAE, Universidade de Santiago de Compostela, 15782 Santiago de Compostela, Spain.

E-mail: camilo@usal.es

Abstract. In this paper, we present a target system to produce protons from laser plasma interactions at high repetition rates. In order to increase the number of available shots needed for the foreseen application, the target system will help to improve the reproducibility and stability of the proton source. We describe the experiments that we will carry out with the 50 TW (1.2J, 25 fs, 10Hz) laser at the Laser Laboratory for Acceleration and Applications (L2A2) at the University of Santiago of Compostela to study the production of radioactive isotopes for medical imaging such as PET. We describe the multi-shot target that we have developed and the related diagnostic to characterize the focal spot and the relative distance between the target and the focus. The multi-shot target will permit the use of thin foils or nano-structured targets with repetition rates relevant to our high power laser system.

1. Introduction

Laser acceleration of ions by ultrashort high power laser system has been a topic of intense research over the last two decades [1, 2, 3]. The production of MeV ion bunches from the interaction of these laser pulses with solid targets is attractive for a wide range of applications such as ion beam therapy[4], fast ignition [5], radiography [6] and production of radio isotopes for nuclear medicine [7, 8].

The physical mechanisms behind the acceleration processes such as the Target Normal Sheath Acceleration (TNSA), Radiation Pressure Acceleration (RPA) and others are well established [3] and several proofs of principle exist. All of these mechanisms have limited efficiency and while some of these mechanisms have a favorable energy scaling for higher laser powers, the main bottleneck towards applications is the average power of the sources [9, 10]. The average power of the source is related to the energy of the ions produced on each shot and the number of shots per second that can be achieved. This last problem is directly related to the target system and the laser repetition rate to be used. A good target system for applications should solve the problem of reproducibility, stability of the source and the need for a large number of shots in the same target[11].

IOP Conf. Series: Journal of Physics: Conf. Series **1079** (2018) 012009 doi:10.1088/1742-6596/1079/1/012009

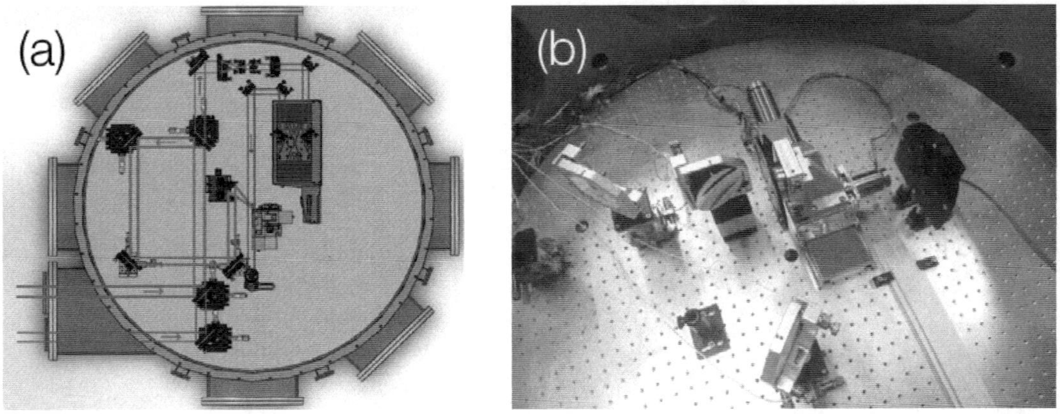

Figure 1: (a) Sketch of the experimental setup. The beam at the bottom of the chamber is an alignment beam, the other beam entering the chamber is the main laser beam . (b) Picture of the focusing OAP and the target system inside the vacuum chamber.

While high power laser systems can provide relatively high repetition rates ($\sim 0.5 - 10Hz$) there are only few examples of target systems that can sustain continuous operation of a proton source at relevant repetition rates [12, 13, 14, 15]. Most of the applications mentioned above require a stable proton source with high average power [11]. The advantages of having targets systems capable of sustaining the repetition rate of the laser are many, such as systematic studies, further optimization of the experimental parameters and ultimately the demonstration of the devised applications.

In this paper we describe the target system developed at the L2A2 facility at the University of Santiago of Compostela to produce ions from laser plasma interactions. The proposed target assembly provide an accurate positioning system of targets for repetition rates (0.1-10 Hz) relevant to high power laser systems for a large number of shots (>1000). The target system should also be capable to support different type of target materials such as thin foils or micron and nano-structured targets made with MEMS technology[**?**]. Such target system could be tested for the application of radio isotope production.

2. Experimental setup

For the ion acceleration experiment we will use a laser system (THALES ALPHA 10/XS) that has two Chirped Pulse Amplification (CPA) stages [16]. In the first stage the pulse is amplified to $E = 1$ mJ at 1 KHz and compressed to 35 fs where a Cross polarization wave (XPW) module [17, 18] is used to enhance the contrast of the pulse. The contrast achieved with this setup is larger than 10 orders of magnitude and will play an important role in the interaction with solid targets. After improving the contrast, the pulse goes through a second CPA amplification stage to reach a final energy of $E = 1.2$ J and after compression a temporal FWHM of 25 fs with a repetition rate of 10 Hz.

The compressor is connected to the experimental chamber through a 8 m vacuum beam line. The laser beam has a 50 mm diameter and enters into the experimental chamber parallel to a diode laser (Z-Laser, 40 mW, 685 nm) enlarged with a telescope to match the laser. We will refer to this beam as the alignment beam. Figure 1 shows the experimental setup. We use a 45° Off Axis Parabola (OAP) with an reflective focal length of 89.28 mm (F/1.8) to produce a 4 μm

IOP Conf. Series: Journal of Physics: Conf. Series **1079** (2018) 012009 doi:10.1088/1742-6596/1079/1/012009

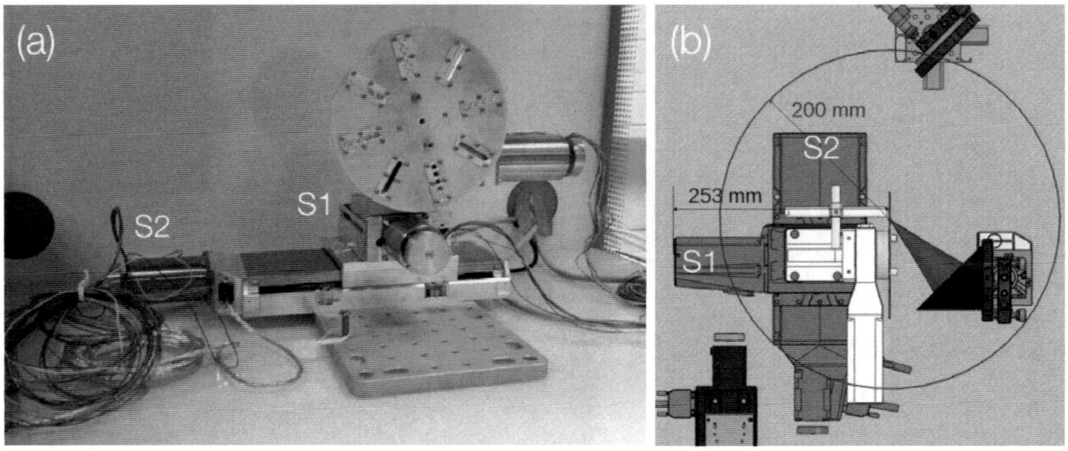

Figure 2: Motor system for target positioning and refreshing.

focal spot which corresponds to an intensity of 4×10^{20} W/cm^2. A leak from one of the mirrors is frequency doubled with a BBO crystal to produce a probe beam which can be delayed from the main beam and that passes parallel to the target surface.

3. Target for multi-shot

The most important challenge for the design of the target system is to refresh the target material at the laser focus with the required accuracy and velocity to sustain a high repetition rate. Our target system consist of a positioning system, a holder for the targets and a diagnostic system to measure the focal spot and to align the target relative to the laser focus. The design of the system takes into account the experimental constrains of the laser focusing and the diagnostics of the interaction.

The positioning system is the most critical part because it must align the target in focus to achieve the highest intensity on target and keep the alignment for all the targets in the disk that holds the targets. The Rayleigh length defines the allowed error and in our case is 30 μm. The positioning system is made of an arrangement composed by one rotating stage (PiMicos, DT-65N) and two linear stages (PiMicos, PLS-85 and LS-110) all stacked in a single assembly as shown in Fig. 2(A). The rotating stage holds a disk where the target foils are located, and it can achieve a rotation speed of 15°/s in high vacuum. This stage is held by an L-shape piece to the first linear stage (S1) which controls the target position relative to the focal plane. This stage is attached on top of the second linear stage (S2) which is used to translate the disk and change the radius where the laser impacts.

We have designed different holders for the target system. The first one shown in Fig. 3(b) is a disk with 8 slots to hold up to 40 different targets in the disk. A second target made with a 3D printer can hold 9 slots each one with 6 targets (Fig.3(b)) and third holder will be used to hold 6" silicon wafer used in the construction of micro structured targets with the MEMS technology [14]. These micro structured targets have shown to enhance the energy absorption and the final energy of the accelerated protons [19, 20].

IOP Conf. Series: Journal of Physics: Conf. Series **1079** (2018) 012009 doi:10.1088/1742-6596/1079/1/012009

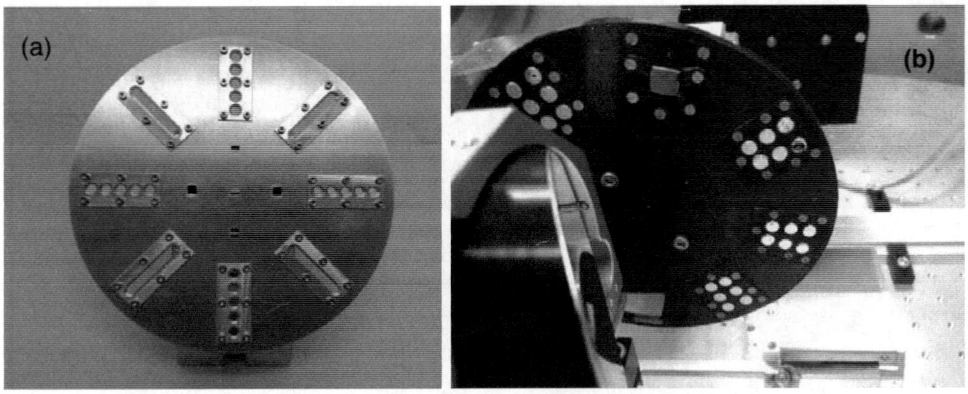

Figure 3: (a) Disc holder for targets. Each circle holds one target, strips of targets can also be used with this holder. (b) Holder for multi-shot operation produced with 3D printer.

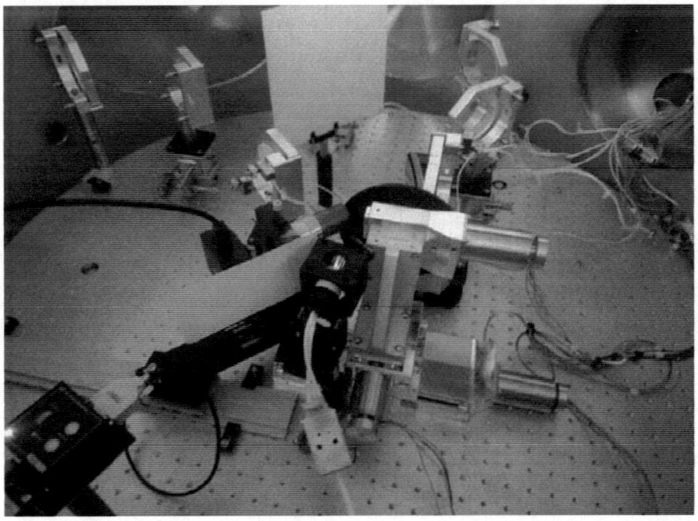

Figure 4: Experimental setup of the diagnostic tool showing the microscope objective

4. Diagnostic of the focus and target positioning

In order to characterize and to measure the spot size of the laser after the OAP we use a microscope objective (Mitutoyo, M Plan Apo NIR 50x) located behind the target in the axis defined by parabola and the visible alignment beam as shown in Figure 4.

IOP Conf. Series: Journal of Physics: Conf. Series **1079** (2018) 012009 doi:10.1088/1742-6596/1079/1/012009

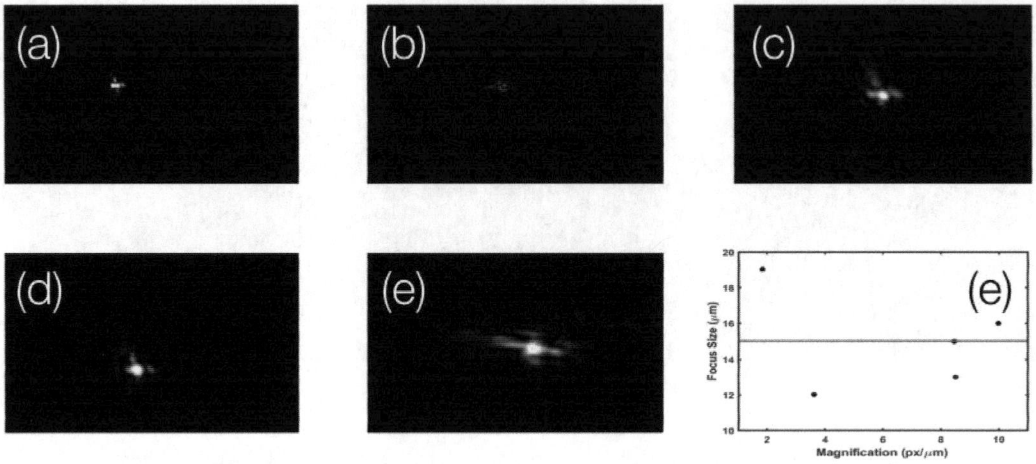

Figure 5: Focal spot of the alignment beam for different magnifications/camera (β) resolutions. (a)β=3.62, (b) β=8.48, (c) β=8.52 (d) β=10.00, (e) Calculated mean value for the focus:15 ± 2 μm

We place the microscope objective and use the alignment laser described before to image the focal spot size with a linear stage with 0.05 μm step size (Thorlabs, MTS25/M-Z8). Once the microscope objective is in place, we measure the spot size with a CMOS camera (5 Mpx, Mightex) for different magnifications, as shown in Figure 5. The measured spot size is 15 ± 2 μm mean value for the alignment beam (Fig. 5(e)), which has larger divergence than the infrared beam. We make use of this setup to position the target surface in the best focus. We do this using two complementary techniques. The first one is imaging the rear side of the target and the other one is the speckle technique.

The first technique works in the following way: Once the laser spot size is characterized and the position of the microscope objective fixed to image the focus of the laser, we place the target holder in front of the laser and look at the rear side of targets using the microscope objective and a white LED light. As the incidence angle of the laser in the target is 45°, we only observe a line in focus in the camera. When the central part of the rear side image in the camera is in the objective focal plane as shown in Figure 6, that surface is at the position of the laser focus. Therefore we just need to use the motor S1 to move the target exactly by the width of the target disk to ensure that the laser focus matches the surface of the target facing the laser. We can use this technique to put in focus up to 10 targets places and record the motor positions for the best focus on target in each of these 10 targets. After we have done this we remove the microscope objective before the shots.

IOP Conf. Series: Journal of Physics: Conf. Series **1079** (2018) 012009 doi:10.1088/1742-6596/1079/1/012009

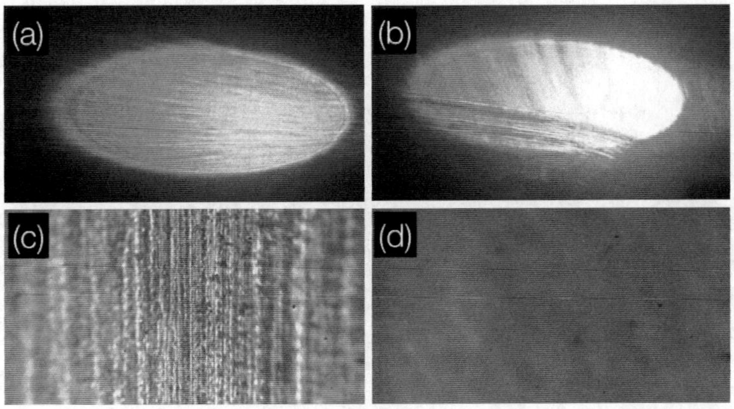

Figure 6: (a) Speckle image of the target on focus. (b) Speckle image of target 100 μm after focus. (c) Rear side image of a thin foil of copper used to test the technique at the best target position. (d) The same copper foil rear side when the target is off focus

While this technique is enough to position the target on focus, we have verified that it matches with the best target position characterized with the speckle. Figure 6 shows the reflection of the alignment beam in the surface of the target. As the target approaches to the focus, the reflection of the alignment beam changes from a rough reflection (Fig. 6(b)) to a smooth reflection (Fig. 6(a)) on focus. The back lighting and the speckle techniques provide very similar positioning of the target at the laser focus. Both techniques have some advantages and problems and we use both complementary. The back lighting technique is useful to characterize the main laser focus but the microscope objective and the camera need to be removed before the high power laser can be used because of the required vacuum. On the other hand, the speckle technique uses a reflection which can be guided to a screen outside of the chamber .

In the Figure 7 an image of the target holder after the laser impact is shown. In each of the target position the laser destroy the target foil. The size of the damage is similar in all used positions which is an indication that the laser intensity has been the same in all of them.

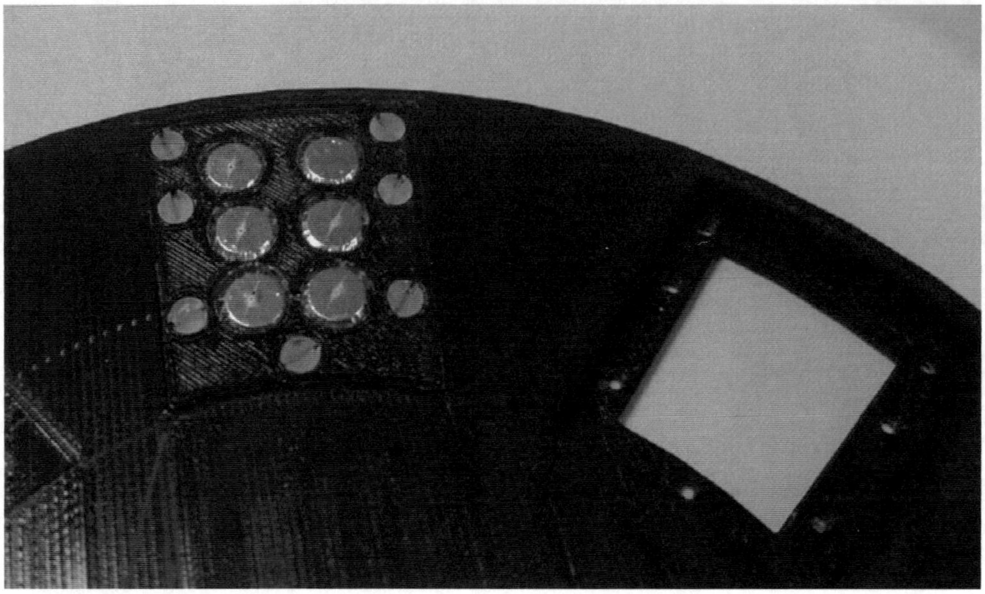

Figure 7: Six targets after the impact with the laser pulse. The position of the motors for the beast focus on target is recorded for each site in the target, this way we guarantee the same intensities on target.

5. Conclusions

Proton acceleration from the interaction of solid targets with high power lasers using the TNSA scheme has been proven to produce ion beams with extraordinary features which could open new applications. In order to explore all these possible applications, it is necessary to design a source with high average power. Such source should have a target system which can work using the repetition rate available from the laser system. In this paper, we have introduced a multi-shot target and an alignment system which would allow us to have a large number of shots without opening the experimental chamber. The final objective is to obtain a multishot target to perform systematic studies and to improve the stability and reproducibilty of the source. The rotating system presented here uses a rotating disk to accomodate > 1000 targets. Also we have verified that the backlighting and the speckle method coincide in the determination of the best overlap between the target and the laser focus which allow us to use them in different situations.

Acknowledgments

This work has been supported by Spanish Ministry of Industry within the framework of the RETOS project (RTC-2015-3278-1) and by Xunta de Galicia and EU, LASERPET project (2013-AD009.01). Camilo Ruiz acknowledges the MINECO project FIS2016-75652-P.

IOP Conf. Series: Journal of Physics: Conf. Series **1079** (2018) 012009 doi:10.1088/1742-6596/1079/1/012009

References

[1] Daido H, Nishiuchi M and Pirozhkov A S 2012 *Reports on Progress in Physics* **75** 056401–72
[2] Schreiber J, Bolton P R and Parodi K 2016 *Review of Scientific Instruments* **87** 071101
[3] Macchi A, Borghesi M and Passoni M 2013 *Reviews of Modern Physics* **85** 751–93
[4] Linz U and Alonso J 2007 *Physical Review Special Topics - Accelerators and Beams* **10** 487–8
[5] Roth M, Cowan T E, Key M H, Hatchett S P, Brown C, Fountain W, Johnson J, Pennington D M, Snavely R A, Wilks S C, Yasuike K, Ruhl H, Pegoraro F, Bulanov S V, Campbell E M, Perry M D and Powell H 2001 *Physical Review Letters* **86** 436–39
[6] Mackinnon A J, Patel P K, Borghesi M, Clarke R C, Freeman R R, Habara H, Hatchett S P, Hey D, Hicks D G, Kar S, Key M H, King J A, Lancaster K, Neely D, Nikkro A, Norreys P A, Notley M M, Phillips T W, Romagnani L, Snavely R A, Stephens R B and Town R P J 2006 *Physical Review Letters* **97** 045001–5
[7] Amato E, Italiano A, Margarone D, Pagano B, Baldari S and Korn G 2016 *Nuclear Inst. and Methods in Physics Research* **A 811** 1–5
[8] Lefebvre E, d'Humières E, Fritzler S and Malka V 2006 *Journal of Applied Physics* **100** 113308–11
[9] Borghesi M 2014 *Nuclear Inst. and Methods in Physics Research* **A 740** 6–9
[10] Fuchs J, Antici P, d'Humières E, Lefebvre E, Borghesi M, Brambrink E, Cecchetti C A, Kaluza M, Malka V, Manclossi M, Meyroneinc S, Mora P, Schreiber J, Toncian T, Pépin H and Audebert P 2005 *Nature Physics* **2** 48–54
[11] Prencipe I, Fuchs J, Pascarelli S, Schumacher D W, Stephens R B, Alexander N B, Briggs R, Büscher M, Cernaianu M O, Choukourov A, De Marco M, Erbe A, Fassbender J, Fiquet G, Fitzsimmons P, Gheorghiu C, Hund J, Huang L G, Harmand M, Hartley N J, Irman A, Kluge T, Konopkova Z, Kraft S, Kraus D, Leca V, Margarone D, Metzkes J, Nagai K, Nazarov W, Lutoslawski P, Papp D, Passoni M, Pelka A, Perin J P, Schulz J, Smid M, Spindloe C, Steinke S, Torchio R, Vass C, Wiste T, Zaffino R, Zeil K, Tschentscher T, Schramm U and Cowan T E 2017 *High Power Laser Science and Engineering* **5** 483–31
[12] Noaman-ul Haq M, Ahmed H, Sokollik T, Yu L, Liu Z, Yuan X, Yuan F, Mirzaie M, Ge X, Chen L and Zhang J 2017 *Physical Review Accelerators and Beams* **20** 041301–6
[13] Gao Y, Bin J, Haffa D, Kreuzer C, Hartmann J, Speicher M, Lindner F H, Ostermayr T M, Hilz P, Rösch T F, Lehrack S, Englbrecht F, Seuferling S, Gilljohann M, Ding H, Ma W, Parodi K and Schreiber J 2017 *High Power Laser Science and Engineering* **5** e12–5
[14] Spindloe C, Arthur G, Hall F, Tomlinson S, Potter R, Kar S, Green J, Higginbotham A, Booth N and Tolley M K 2016 *Journal of Physics: Conference Series* **713** 012002–9
[15] Booth N, Clarke R, Heathcote R, Neely D, Pattathil R, Rusby D, Spindloe C, Symes D, Tolley M and Tomlinson S 2014 *SPIE Optical Engineering + Applications* ed Bell P M and Grim G P (SPIE) pp 921107–9
[16] Mourou G A, Tajima T and Bulanov S V 2006 *Rev. Mod. Phys.* **78**(2) 309–71 URL `https://link.aps.org/doi/10.1103/RevModPhys.78.309`
[17] Jullien A, Albert O, Burgy F, Hamoniaux G, Rousseau J P, Chambaret J P, Augé-Rochereau F, Chériaux G, Etchepare J, Minkovski N and Saltiel S M 2005 *Opt. Lett.* **30** 920–22 URL `http://ol.osa.org/abstract.cfm?URI=ol-30-8-920`
[18] Jullien A, Chen X, Ricci A, Rousseau J P, Lopez-Martens R, Ramirez L P, Papadopoulos D, Pellegrina A, Druon F and Georges P 2011 *Applied Physics* **B 102** 769–74 ISSN 1432-0649 URL `https://doi.org/10.1007/s00340-010-4201-9`
[19] Blanco M, Flores-Arias M, Ruiz C and Vranic M 2017 *New Journal of Physics* **19** 033004
[20] Fedeli L, Formenti A, Cialfi L, Sgattoni A, Cantono G and Passoni M 2018 *Plasma Physics and Controlled Fusion* **60** 014013 URL `http://stacks.iop.org/0741-3335/60/i=1/a=014013`

IOP Conf. Series: Journal of Physics: Conf. Series **1079** (2018) 012010 doi:10.1088/1742-6596/1079/1/012010

Hydrogen resist lithography and electron beam lithography for fabricating silicon targets for studying donor orbital states

E. Crane, A. Kölker, T. Z. Stock, N. Stavrias, K. Saeedi, M. A. W. van Loon, B. N. Murdin, N. J. Curson

London Center for Nanotechnology, University College London, Gower St, London, UK

E-mail: e.crane@ucl.ac.uk

Abstract. Recently, phosphorous structures in silicon have been of interest theoretically and experimentally due to their relevance in the field of quantum computing. Coherent control of the orbital states of shallow donors in silicon has been demonstrated in bulk doped samples. Here we discuss the fabrication techniques required to 1) obtain patterned two dimensional dilute sheets of impurities in silicon of controlled doping densities 2) get them to act as targets for a terahertz laser. Scanning tunnelling microscope hydrogen lithography enables patterning of impurity features in silicon with a resolution from 1nm to tens of nm. Molecular beam epitaxy is used for a protective thin-film crystalline silicon growth over the impurity sheet. Electron beam lithography coupled with reactive ion etching allows features from tens to hundreds of microns to be etched into the silicon with 10 to 20nm resolution. The experimental readout is achieved via illumination of the silicon target by terahertz light and subsequent electrical detection. The electrical signal comes from coherent and non-linear excitations of the impurity electrons. This detection technique enables the precision condensed matter samples to remain intact after exposure to the free electron laser pulse.

1. Introduction

Investigating the properties of phosphorous impurities in silicon is crucial to forward the attemps which are being made to create entangling gates utilising the spins or atomic orbitals of the donor electrons[1][2][3]. While control of the phosphorous atomic orbital states by terahertz light has been demonstrated in 3D bulk doped Czochralski crystal grown silicon wafers [4][Chr], their control has not yet been demonstrated in two dimensions (2D). Deterministic placement of donors is only possible to a sufficiently a high degree of precision with scanning tunnelling microscope (STM) hydrogen lithography, which takes place in 2D (it can be layered to create deterministic 3D structures). For this reason, understanding 2D randomly doped impurity samples is a key stepping stone to deterministic positioning of entangling gates. Here we present the method developed to fabricate 2D dopant layer samples [5].

Coherent control of phosphorous Rydberg states in silicon has been demonstrated with optical and electrical detection [6][7].

The fabrication techniques used to grow the phosphorous 2D layer will first be presented. A range of lithographic techniques are used to transfer patterns to the surface of the samples: STM, hydrogen lithography and molecular beam epitaxy [11]. Secondly, the

IOP Conf. Series: Journal of Physics: Conf. Series **1079** (2018) 012010 doi:10.1088/1742-6596/1079/1/012010

techniques for accessing the buried 2D layer will be presented: electron beam lithography (EBL), reactive ion etching (RIE) and a hydrofluoric acid dip. Lastly, the method for electrical detection will be discussed.

The experiments are led collaboratively by the London Centre for Nanotechnology at University College London where the targets are produced and characterised, and by the University of Surrey and Radboud University in the Netherlands where the Free Electron Laser for Infrared eXperiments (FELIX) is used to perform the experiment [3].

2. Hydrogen lithography for the delta layer preparation

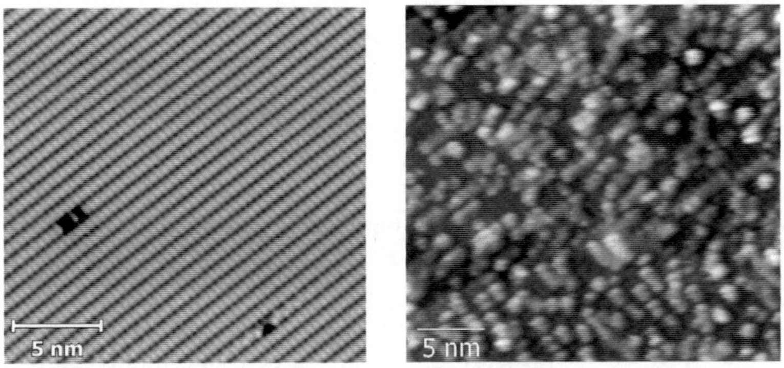

Figure 1. (left) Filled state STM image showing a clean Si(001)-2X1 surface. Silicon dimer rows run diagonally across the image and are interrupted by a low density of native defects. Pictured here are a 2+1 dimer vacancy defect, and a c-defect due to dissociation of water[12]. Image acquired with -2.0 V, 2 nA. Image courtesy of Taylor J.Z. Stock, London Centre for Nanotechnology, University College London. (right) Empty state STM image of an annealed Si(001)-PH3 surface. The silicon is initially exposed to a monolayer saturation coverage of PH3, and then annealed at 350 C for 5 minutes. This anneal allows P atoms to incorporate into the surface lattice, and ejected silicon atoms are observed to form incomplete and isolated dimer rows on top, running perpendicular to the directions of the underlying surface dimer rows. Image acquired with +2.0 V, 2 nA.

In ultra-high vacuum (UHV), a dilute 2D layer of P or As donors is incorporated into the clean silicon crystal lattice. It's these 'isolated' dopants on which the terahertz experiments are performed. To establish a good metallic contact to the dilute 2D layer, highly conductive metallically doped phosphorous pads of 0.2 x 0.2 um are patterned on the same plane using STM hydrogen lithography. These 2D layers are overgrown with a protective layer of silicon. The metallic in plane pads are contacted using a process described in the next section, which utilises the technique of EBL.

The initial wafers are bulk n-type (10^{14} dopants per cm^3 and above) Czochralski grown silicon. Deep etched orientation markers are transferred into the silicon (001) wafers via optical lithography to determine the location of the STM tip on the sample surface. All the wafers are in the [001] direction and nominally flat. They are only handled with ceramic tweezers in order to avoid contamination. They are then diced into 2 mm x 9 mm slices to fit into the Omicron variable temperature STM plates made of tantalum, molybdenum or ceramic.

After introduction into UHV in the preparation chamber, the sample holder is heated to 300°C, at which point the n-type silicon becomes conductive. Direct current is then passed through them until they reach temperatures of 600°C (confirmed by infra-red pyrometry), all the while keeping the pressure of the chamber below 10^{-11} mBar. They are left to

Figure 2. Hydrogen passivated silicon where two 0.2um x 0.2um squares have been desorbed with the STM tip. Desorbtion parameters: bias of 7.0 V, setpoint current of 1.0 nA, dose of 0.2 mC/cm and tip speed of 50.0 nm/s. Image acquired with -5V, 0.05nA. Image courtesy of A. Kölker, London Centre for Nanotechnology, University College London.

outgass overnight which removes the water absorbed onto both the sample and sample holder when they were in air [11].

The samples are flashed, ie. successively heated to temperatures of 1150°C four or five times, for up to 20 seconds at a time. This removes the native oxide on the surface of the silicon, as well as the impurities it accumulated before entering ultra-high vacuum [11]. The wafer is then ready to be processed in the STM chamber of base pressure 10^{-11} mBar.

The metallic tip of the STM is brought within a few angstroms of the conductive sample surface. Applying a bias between the two induces a quantum mechanical tunnelling current to flow. While the tip is raster scanned across the surface of the sample, it's height is altered in order to maintain the tunnelling current constant. Recording the positions of the tip at successive locations across the sample generates a map of the topography of the surface.

In order to ensure the sample is clean, randomly chosen small portions of the surface is scanned. Contamination may come from the tip or the sample holder (even in infinitesimal amounts, tungsten or nickel create dimer vacancy defects in the silicon surface) [12]. When the silicon is pristine, dimer rows are visible on the surface as in figure 1 (left).

Phosphine gas (PH3) is released onto the surface. The main sequence of dissociation is PH2+H, PH+2H, and P+3H, with the secondary sequence being PH+H to P+2H [13]. Phosphorous atoms are randomly distributed on the surface at a low coverage, the exact density of which can be determined by measuring the pressure of the gas in the chamber at the time of the gas exposure.

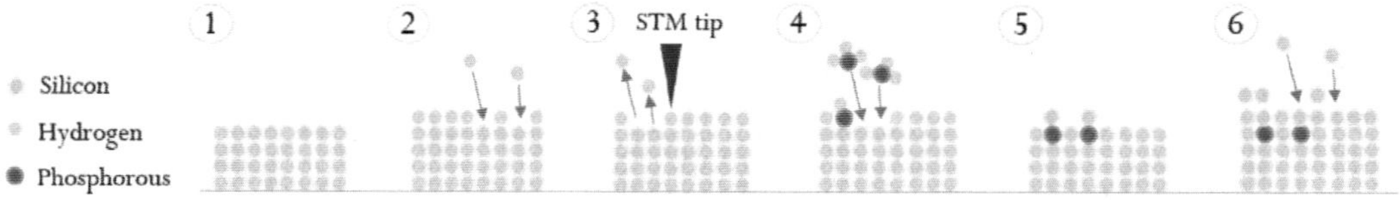

Figure 3. Process used in ultra-high vacuum scanning tunneling microscope chamber to carry out hyrdogen lithography on clean silicon wafers, dose with phosphine gas and overgrow with silicon.

The sample is heated to $\sim$ 350°C for two minutes, incorporating all the phosphorous into the silicon. This process has taken place when bright protrusions (elevated atoms) are seen on the STM image which indicate that the phosphorous impurities have ejected the silicon atoms below them, which have then accumulated on the surface [11], see figure 1 (right).

IOP Conf. Series: Journal of Physics: Conf. Series **1079** (2018) 012010 doi:10.1088/1742-6596/1079/1/012010

The process of hydrogen lithography is used to define features on the silicon surface, as can be seen in figure 2. The sample is heated to 325°C with direct current passing through it. Exposure to hydrogen gas cracked by a hot tungsten rod results in a atomically thick layer of hydrogen covering the silicon. Applying a high voltage 8V and high setpoint currents of up to 3 nA between the STM tip and the sample creates enough energy to locally break the bond between the individual hydrogen atom and its supporting silicon [14]. The resulting half filled orbital protruding into the vacuum is called a dangling bond and is far more reactive than the surrounding robust hydrogen resist. Individual atoms of hydrogen can thus be removed to create atomically precise patterns [15]. Applying higher voltages and tip movement speeds enables larger areas to be desorbed, albeit with a lower precision. The technique is applicable for total areas of tens of nm to tens of microns and has also been demonstrated to enable placement of dopants to within 1nm precision [15].

Feedback controlled lithography has been developed and demonstrated to achieve high resolution nanopatterning [16]. The hydrogen resist can also be used as a hermetic layer to protect surfaces prepared in UHV from ambient conditions [17]. Placing two hydrogen passivated surfaces in contact with one another creates Van der Waals bonds between the hydrogen atoms and can protect the surfaces from being oxidised or contaminated for weeks [17]. It is also possible to identify previously positioned dopants underneath the hydrogen layer using the STM [18][19].

When exposed to phosphine gas only the exposed areas are reactive, not the ones protected by the hydrogen. The phosphine gas dissociates on the exposed areas, and remains local during the incorporation which is done by heating as previously discussed.

Molecular beam epitaxy enables a protective layer of silicon to be grown above the incorporated impurities. The sample is heated to 250°C by direct current while a silicon sublimation cell is heated to over 1000°and aimed at it. Single silicon atoms land on the surface and construct a thin crystalline film, following the base crystal structure of the sample. This technique can be used to grow tens to hundreds of nm of crystal on surfaces. Typically, the samples are encapsulated with 15 nm to 30 nm of epitaxial silicon. The deposition rate is of ≈ 0.125 nm/min.

A schematic of the entire process taking place in ultra-high vacuum can be seen in figure 3.

3. Electron Beam lithography for accessing the delta layer

In order to make electrical contacts to the buried nanostructures, EBL (or e-beam) is used in conjunction with RIE, and metal evaporation. The aim is to create small holes (vias), 70 nm in diameter and 60 nm deep into the silicon, which traverse the 0.2 x 0.2 um square patches of metallically doped phosphorous 15 nm to 30 nm below the surface. The sample is then exposed to a hydrofluoric acid dip and the vias are filled with aluminium to create metallic vias which maximises the surface area in contact with the phosphorous layer.

A thin organic film of resist coating the sample is exposed to a tightly focused and precisely controlled electron beam. The mask is directly defined by scanning the electron beam. The energy delivered to the resist in these regions chemically alters it [20]. When the coated sample comes into contact with a solvent (developer), the altered or non-altered regions are removed depending on if the resist is positive or negative respectively. EBL can create features with 10 nm - 20 nm resolution using polymer resist due to the size of the polymer molecules and the electron scattering during exposure [21]. The advantage of EBL over photo-lithography stems from the shorter wavelength of accelerated electrons compared to the wavelength of ultraviolet light.

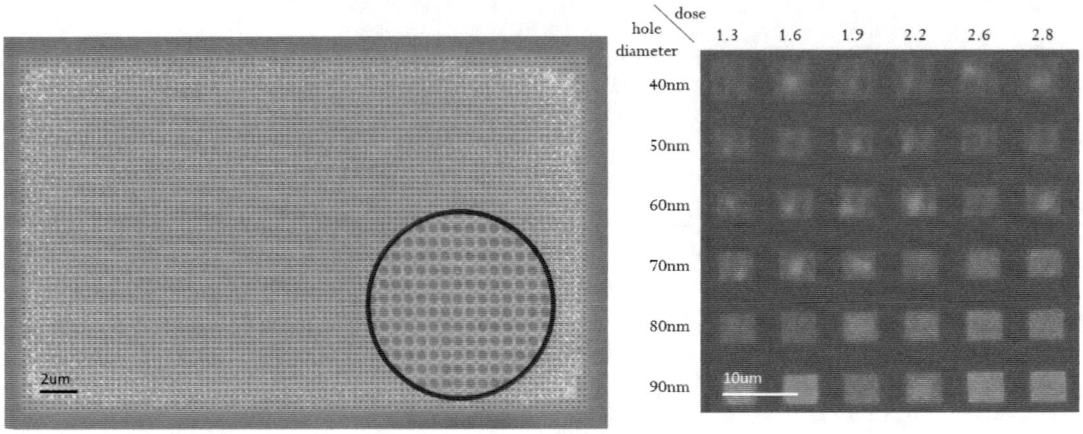

Figure 4. Scanning Electron Microscope (SEM) images of holes defined in resist using EBL. Beam Power used for SEM imaging and for EBL hole definition: 10 keV. Working distance: 10 mm. Doses are factors of the standard dose: 10 mu A/cm^2. (left) Etched holes of 70 nm diameter, 60 nm deep in the silicon sample. The black circle is a zoom on the holes. (right) A series of test holes etched into a test silicon sample.

The small, circular and stable electron beam is provided by an integrated Scanning Electron Microscope (SEM) [22]. To achieve high quality beam parameters UHV is used. A field emission source is used to create the electrons which scan the sample and expose the resist and is composed of one or two electrodes which extract and accelerate electrons pulled from a filament [20]. The beam brightness (equivalent to intensity in conventional optics) should be maximised to minimise exposure time, but in practice high resolution exposures are only achieved with lower intensities. Thinning the resist can improve the resolution of the features [21].

The pattern is written using software which is specific to the EBL equipment. As can be seen on figure 4, EBL has been used to create a mask containing circles in the resist. The resist is exposed to a beam of electrons of around 200 pA. The final dose corresponds to the amount of charge per unit area that the resist receives [20], it is obtained by multiplying the dose factor with the standard dose of 110 mu A/cm^2. The optimal dose found here was for holes with a diameter of 70 nm with a dose factor of 2.2, from which is the image displayed on figure 4 (left) was taken. The insides of the circles in pattern on the resist are then removed during the development stage. The resist-covered regions remain protected during further processing (until its removal). After cleaning, the sample is introduced into the RIE, which uses an etchant gas to remove the silicon in the exposed regions of the pattern.

The sample is covered in resist again, and contact pads are written onto the resist, aligned on the array of holes. After the resist has been developed, 150 nm of aluminium is deposited onto the surface. A lift-off removes the resist and covering aluminium around the pads. The silicon is left with aluminium contact pads with deep vias.

Test patterns with different exposure parameters and length scales must be carried out on a test sample beforehand such that the current beam parameters of the machine is accounted for, as is shown in figure 4 (right). The dose depends on the number of incoming electrons required to fully expose the resist and varies relative to the thickness, shape or area of the resist to be exposed.

EBL is also compatible with mono-atomically thick hydrogen, similar to STM hydrogen lithography (which can then have aluminium features grown onto the remaining hydrogen [21]); however, EBL hydrogen lithography is limited to 5 nm - 50 nm resolution.

IOP Conf. Series: Journal of Physics: Conf. Series **1079** (2018) 012010 doi:10.1088/1742-6596/1079/1/012010

4. Electrical detection for non-destructive experimental result readout

The signal which arises due to the illumination of the low doped phosphorous 2D layer by the free electron laser beam can be readout optically or electrically [9]. Electrical detection is a more sensitive method and is therefore better adapted to samples where the phosphorous is confined to two dimensions, compared to the samples where the impurity species is introduced into the silicon during the melt, before the crystal is grown and therefore present in higher quantities per sample volume. Electrical readout, has been developed in silicon for the purposes of terahertz control of donors in silicon [6] and is possible with both 2D and 3D doping.

The 2p0 and 2p+- P states are split by the electron's anisotropic effective mass in silicon. Terahertz light tuned to the exact transition energy (39.19 meV)/ wavelength (31.64 um)/ frequency (9.48 Thz) selectively excites electrons from the insulating low doped phosphorous (located between the conductive 2x3um metallic pads in the 2D layer) from the 1s to 2p+- state. Once excited to this state, it can get excited into the conduction band by collisions with thermal phonons. It is electrically detected when a bias is applied across the sample. The beam size is between 0.02cm to 0.1cm, with cylindrical symmetry and a Gaussian intensity profile. For a 2D sample, this implies that around This technique, called photothermal ionisation spectroscopy (PTIS)[7], requires the sample to be electrically contacted. The mechanisms involved in PTIS are shown in figure 5 (left).

Typically, approximately 4 pA current per electron is detected. A graph of the current detected as a function of the free electron laser wavelength is shown in figure 5 (right), illustrating that it is possible to measure the number of donors excited to a specific state using electrical detection. The wavelengths at which the donor electrons to get from the 1s to the 2p0 and 2p+- states (corresponding to 36.4 um and 31.6 um respectively), are indicated on the graph.

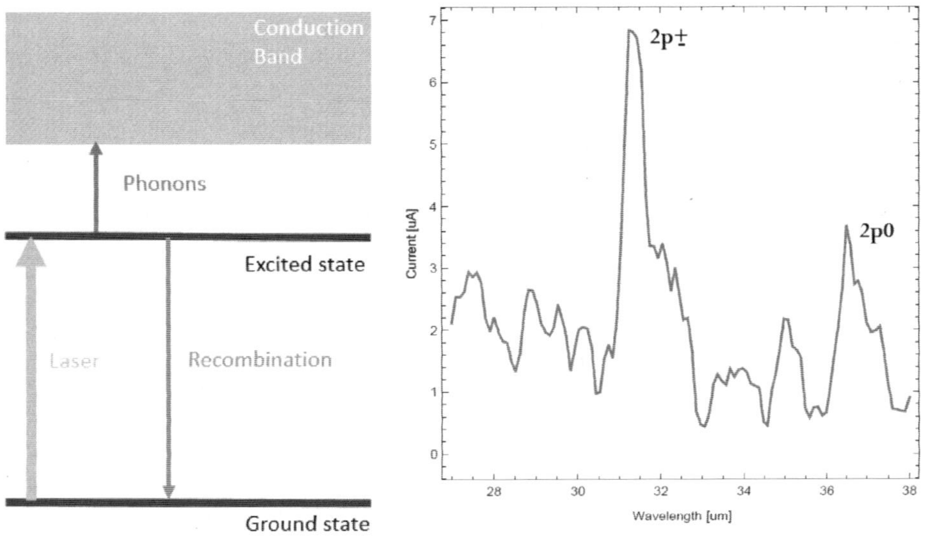

Figure 5. (left) Processes involved in Photo-Thermal Ionisation Spectroscopy. (right) Electrical signal as a function of terahertz laser wavelength in a contacted silicon sample containing a 2D phosphorous doped layer.

The samples can be current biased or voltage biased[6][7]. The picosecond dynamics can be accessed by voltage biasing the sample[6]. Electrical detection enables the samples to not be destroyed during the measurement. Using electrical detection, the homogeneous phase decoherence rate has been demonstrated to be around around 160ps for the 2p+- state of phosphorous in silicon in a silicon sample of concentration 2 . 10^{-14} dopants per cm^3 [9].

Conclusion

Targets containing 2D phosphorous nanostructures can be made in silicon using a range of lithography, etching and overgrowth techniques. STM hydrogen lithography has up to 1 nm resolution and is a delicate process requiring ultra-high vacuum, whereas EBL using polymer resist has tens of nm resolution but can cover much larger areas. These techniques may be applicable in other areas of silicon laser target fabrication, such as those requiring shot power to be retained by the structure of the targets.

Acknowledgements

We gratefully acknowledge the financial support from the UK Engineering and Physical Sciences Research Council [COMPASSS/ADDRFSS, Grant No. EP/M009564/1]. We thank the Stichting voor Fundamenteel Onderzoek der Materie (FOM) for the support of the FELIX Laboratory

References

[1] Stoneham A M, Fisher A J and Greenland P T 2003 *Journal of Physics: Condensed Matter* **15** L447–L451
[2] Kane B E, 1998 *Nature* **393** 133-7
[3] Stoneham A M 2003 *Physica B: Condensed Matter* **340–342** 48-57
[4] Greenland P T, Lynch S A, van der Meer A F G, Murdin B N, Pidgeon C R, Redlich B, Vinh N Q and Aeppli G 2010 *Nature* **465** 1057-61
[5] Matmon G, Ginossar E, Villis B J, Kolker A, Lim T, Solanki H, Schofield S R, Curson N J, Li J, Murdin B N, Fisher A J and Aeppli G 2018 *Phys. Rev. B* **97** 155306
[6] Bowyer E T, Villis B J, Li J, Litvinenko K L, Murdin B N, Erfani M, Matmon G, Aeppli G, Ortega J M, Prazeres R, Dong L and Yu X 2014 *Appl. Phys. Lett.* **105** 021107
[7] Greenland P T, Matmon G, Villis B J, Bowyer E T, Li J, Murdin B N, van der Meer A F G, Redlich B, Pidgeon C R and Aeppli G 2015 *Phys. Rev. B* **92** 165310
[8] Vinh N Q, Greenland P T, Litvinenko K L, Redlich B, van der Meer A F G, Lynch S A, Warner M, Stoneham A M, Aeppli G, Paul D J, Pidgeon C R and Murdin B N 2008 *PNAS* **105** 10649
[9] Tomaszewski P *Jan Czochralski i jego metoda* 2003 (Wrocław–Kcynia: Oficyna Wydawnicza ATUT)
[10] Litvinenko K L, Bowyer E T, Greenland P T, Stavrias N, Li J, Gwilliam R, Villis B J, Matmon G, Pang M L Y, Redlich B, van der Meer A F G, Pidgeon C R, Aeppli G and Murdin B N 2015 *Nat. Comm.* **6** 6549
[11] Tennant D M, Bleir A R and Wiederrecht G 2009 *Handbook of Nanofabrication* (Amsterdam: Elsevier)
[12] Warschkow O, Schofield S R, Marks N A, Radny M W, Smith P V, and McKenzie D R 2008 *Phys. Rev. B* **77** 201305(R)
[13] Ruess F J, Oberbeck L, Simmons M Y, Goh K E J, Hamilton A R, Hallam T, Schofield S R, Curson N J and Clark R G 2004 *Nano Letters* **4** 1969
[14] Warschkow O, Curson N J, Schofield S R, Marks N A, Wilson H F, Radny M W, Smith P V, Reusch T C G, McKenzie D R and Simmons M Y 2016 *J. Chem. Phys.* **144** 014705
[15] Walsh M A and Hersam M C 2009 *Annu. Rev. Phys. Chem* **60** 193
[16] Lyding J W, Shen T-C, Abeln G C, Wang C and Tucker J R 1996 *Nanotechnology* **7** 128
[17] Schofield S R, Curson N J, Simmons M Y, Rueß F J, Hallam T, Oberbeck L and Clark R G 2003 *Phys. Rev. Lett.* **91** 136104
[18] Grey F, Hermansson K 1997 *Appl. Phys. Lett.* **71** 3400
[19] Brazdova V, Bowler D R, Sinthiptharakoon K, Studer P, Rahnejat A, Curson N J, Schofield S R and Fisher A J 2017 *Phys. Rev. B* **95** 075408
[20] Oberbeck L, Curson N J, Hallam T, Simmons M Y and Clark R G L 2004 *Thin Solid Films* **23** 464
[21] Rius Sune G 2008 Electron Beam Lithography for Nanofabrication, Ph.D (Universitat Autonoma de Barcelona)
[22] Masu K and Tsubouchi K 1994 *J. Vac. Sci. Technol. B* **12** 3270

IOP Conf. Series: Journal of Physics: Conf. Series **1079** (2018) 012011 doi:10.1088/1742-6596/1079/1/012011

A setup for micro-structured silicon targets by femtosecond laser irradiation

Nico W. Neumann[1], Tina Ebert[1], Gabriel Schaumann[1] and Markus Roth[1]

1) Technische Universität Darmstadt, Institut für Kernphysik, Detektor und Targetlabor; Schlossgartenstraße 9, D-64289 Darmstadt, Germany

nico.neumann@physik.tu-darmstadt.de

Abstract. We present a process to fabricate micro-structured thin silicon foils with highly light-absorbing properties over a broad wavelength region based on short-pulse laser treatment. We employ this fascinating technique to the fabrication of targets for high-intensity laser-plasma experiments. A polished silicon wafer is processed with high-intensity femtosecond laser pulses resulting in conical spikes within the irradiated region. Height, distance, and shape of these spikes depend primarily on the number, energy, central wavelength and duration of the incident laser pulses, as well as the ambient medium. This method is of great value for the future development of target fabrication. The broad parameter base offers a huge selection regarding shape and dimensions of the resulting structure. It can be included in existing laser operations within the manufacturing chain while offering a high degree of customisability. Within the reach of higher repetition rates of high-power laser systems and an increased number of requested targets, this manufacturing method can be scaled while being cost efficient and easily adaptable.

1. Introduction

The need for powerful, compact and reliable high energy particle and radiation sources is an ongoing key motivation for new techniques and materials investigated within the field of high power laser plasma science [1-3]. For these experiments the coupling efficiency of laser energy into the interaction is essential. To improve the conversion efficiency of laser energy to ions, electrons or electromagnetic radiation, the precise tuning of laser and target parameters is required. A series of experiments has shown the influence of nano- and micro-structured targets on high power laser experiments [4-7]. We focus on an improved target design with structural modifications of the front surface, intended to increase absorption and alter the emission of electrons, ions, electromagnetic radiation.

Here we produce a structured surface on silicon using a laser driven ablation process described in 1998 by Her. et al. [8]. Following this work, a large number of experiments and further development towards photoelectrical and -chemical applications for micro-structured (black) silicon have been demonstrated for different structural parameters and materials. The dramatically increased absorption over a broad wavelength region is explained to be due to an increased surface area with the cones trapping the light, while a modified band gap further increases the absorption range as a result of laser induced dopants in the surface of the structure. To our knowledge, our work is the first time this fabrication technique has been deployed to the generation of targets for high power laser plasma experiments.

IOP Conf. Series: Journal of Physics: Conf. Series **1079** (2018) 012011 doi:10.1088/1742-6596/1079/1/012011

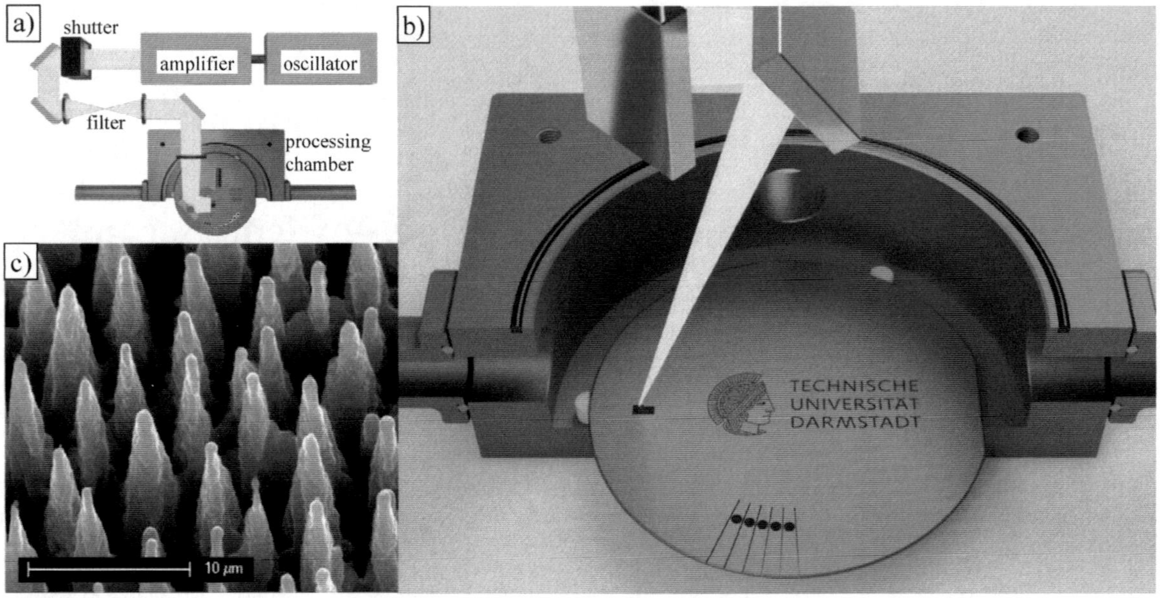

Figure 1: Schematic setup for structuring of silicon. a) shows a simplified overview of the laser system and processing chamber, including the ultrafast shutter and spatial frequency filter. b) is a detailed cross-sectional view of the processing chamber. Precise control of the lateral position of the laser focus on the wafer surface is possible using the two-dimensional galvanometric mirror system. The chamber can be sealed and filled with gaseous and liquid media to influence the general shape of the structure. c) shows a scanning electron microscope image of the targets presented in this work. Conical spikes evolve due to the irradiation of silicon with ultrashort laser pulses in a sulphur hexafluoride ambient.

2. Target Fabrication System

The schematic setup used for the production of structured silicon targets is depicted in Figure 1a). A two-staged Ti:sapphire laser system with a central wavelength of 800 nm, a pulse duration of 100 fs and a repetition rate of 5 kHz is used as driving source for the micro-structuring of silicon. The pulse energy of the laser system can be adjusted continuously in between 10 µJ and 200 µJ. The spatial beam profile is improved by spatial frequency filtering and nearly Gaussian. Using a f=250 mm lens, the laser radiation is focused onto the target surface. Displacement of the focus is possible using a two-dimensional galvanometric mirror system.

Targets described here were produced from mono-crystalline silicon samples with an initial thickness of 25 µm and a ⟨100⟩ surface orientation. These thin membranes were etched in a thicker wafer (600 µm thickness) and placed inside a sealed processing chamber. The chamber can be filled with different gaseous and liquid media. Here sulphur hexafluoride (SF_6) with a pressure of 800 mbar is used. The initially polished silicon surface is illuminated by several hundred femtosecond laser pulses, which results in the ablation of material. Surprisingly, no bulk removal of material will occur, but conical spikes of different height, distance and shape will evolve, depending on the laser fluence, number of incident pulses, central wavelength, pulse duration and ambient medium. It is commonly assumed that these structures are generated by an interference of incoming and scattered light and the excitation of surface plasmon-polaritons. The ultrafast laser-solid interaction causes the electrons in the lattice to heat rapidly. Depending on the pulse intensity this results in melting and rapid solidification of the surface or in an ablation and thereby removal of surface atoms. Respective studies show that processing in air or vacuum creates blunt spikes on a micrometer scale, whereas halogen containing gases result in sharper spikes [9]. Structuring in liquid media will produce smoother variations with a smaller scale length [10].

IOP Conf. Series: Journal of Physics: Conf. Series **1079** (2018) 012011 doi:10.1088/1742-6596/1079/1/012011

3. Experimental Campaign

A comprehensive experiment has been realized at the Central Laser Facility located at the STFC Rutherford Appleton Laboratory, UK. Target Area Petawatt with the high power VULCAN OPCPA laser system offers ideal conditions to explore the characteristic of the novel target design presented here.

a) b)

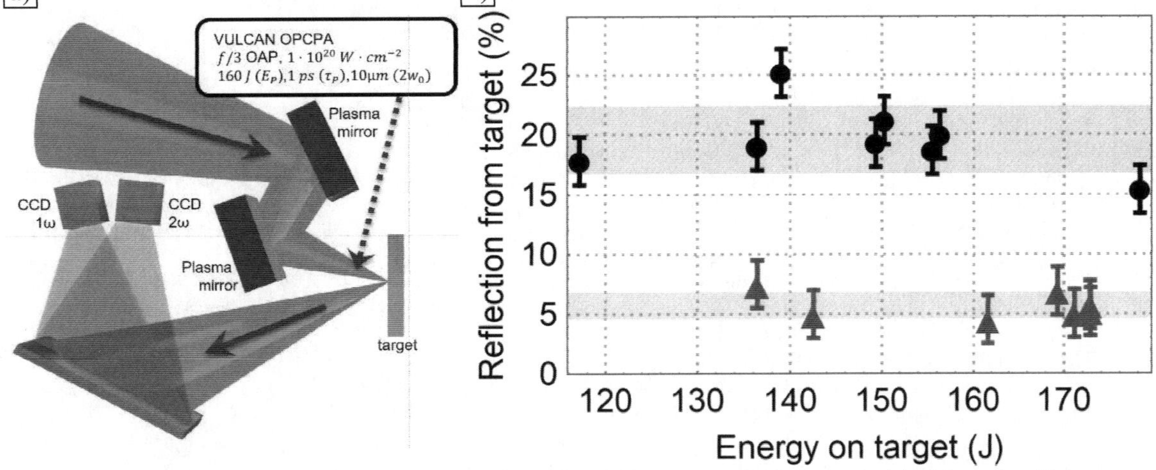

Figure 2: The experimental setup for the campaign at the VULCAN laser system is shown in a). A high intensive, high contrast laser pulse is focussed on the target front. Radiation from the target front is collected on a scattering screen, that is observed by two wavelength selective cameras (fundamental 1053 nm and second harmonic 527 nm radiation). In the campaign, structured targets are compared to flat targets. b) shows the reflection of light from the target front for flat (black) and micro-structured (red) targets. We observe a decrease in reflectivity by a factor of three, using micro-structured silicon targets. The energy coupling to the target is thereby increased.

To observe effects of the front structure interacting with the highest intensity of a laser pulse, the temporal intensity contrast of the laser system is critical, as otherwise the fine structure will be preheated and thereby altered before the main pulse arrives. For this experiment a double plasma mirror system is employed to decrease the pre-pulse intensity below the laser-induced damage threshold for silicon. The pulse energy delivered on target is 160 J with a pulse length of 1 ps, resulting in a peak intensity of $2 \cdot 10^{20}$ $\mathrm{Wcm^{-2}}$ for a focus diameter of 10 μm. Incident angle to target normal is 20°. Thereby it is possible to collect light escaping the target front on an energy calibrated scattering screen and gain an absolute measure for the energy reflected from target.

We observe a strong transition in fundamental reflectivity and second harmonic (SH) emission when comparing flat to micro-structured silicon targets. For 1053 nm we observe a decrease by a factor of 3 (20 ± 3 % to 6 ± 1 %) in reflected laser light of fundamental wavelength (compare Figure 2b). In addition, SH emission is reduced by a factor of 10 (1.0 ± 0.5 a.u. to 0.10 ± 0.05 a.u.). The reduction in light escaping the front surface indicates that the energy coupling to the target is enhanced significantly, while the surface energy flux is decreased due to the increase in interacting surface area. This observation is consistent with previous studies using structured surface targets. With an increased absorption of energy to the target, changes to laser-driven particle and radiation generation is expected and observed. However, the thorough description of the experimental results would go beyond the framework of this article.

4. Conclusion

We demonstrate the production of micro-structured silicon targets with highly light-absorbing properties for high power laser experiments. The targets are structured, using ultrafast laser ablation in a sulphur hexafluoride ambient with a straight-forward and robust fabrication system as described. The resulting microstructure consists of conical spikes, that introduce strong light trapping over a broad wavelength region. First experimental results show a great potential of this novel target fabrication technique for high power plasma science, even towards high repetition experiments.

Acknowledgements
This project received financial support by the DFG in the framework of the Excellence Initiative, Darmstadt Graduate School of Excellence Energy Science and Engineering (GSC 1070).

References
[1] Wilks S C, Kruer W L, Tabak M and Langdon A B 1992 *Phys. Rev. Lett.* **69** 1383
[2] Daido H, Nishiuchi M and Pirozhkov A S 2012 *Rep. Prog. Phys* **75** 056401
[3] Macchi A, Borghesi M and Passoni M 2013 *Rev. Mod. Phys.* **85** 751
[4] Klimo O, Psikal J, Limpouch J, Proska J, Novotny F, Ceccotti T, Floquet V and Kawata S 2011 *New J. Phys.* **13**, 053028
[5] Christoforetti G, Londrillo P, Singh P K, Baffigi F, D'Arrigo G, Lad A D, Milazzo R G, Adak A, Shaikh M, Sarkar D, Chatterjee G, Jha J, Krishnamurthy M, Kumar G R and Gizzi L A 2017 *Sci. Rep.* **7** 1479
[6] Lübcke A, Andreev A A, Höhm S, Grundwald R, Ehrentraut L and Schnürer M 2017 *Sci. Rep.* **7** 44030
[7] Zhao Z, Cao L, Cao L, Wang J, Huang W, Jiang W, He Y, Wu Y, Zhu B, Dong K, Ding Y, Zhang B, Gu Y, Yu M Y and He X T 2010 *Phys. Plasmas* **17** 123108
[8] Her T-H, Finlay R J, Wu C, Deliwala S and Mazur E 1998 *Appl. Phys. Lett* **73** 1673
[9] Sheehy M A, Winston L, Carery J E, Friend C M and Mazur E 2005 *Chem. Matter* **17** 14
[10] Liu H, Chen F, Wang X, Yang Q, Bian H, Si J and Hou X 2010 *Thin Solid Films* **518** 18

IOP Conf. Series: Journal of Physics: Conf. Series **1079** (2018) 012012 doi:10.1088/1742-6596/1079/1/012012

Combining two-photon-polymerization with UV-lithography for laser particle acceleration targets

T Abel[1], R Dörr[2], W Krüger[1], R Horstmann[2], G Schaumann[1], M Roth[1] and H F Schlaak[2]

[1] Institut für Kernphysik, Technische Universität Darmstadt, Schlossgartenstr. 9, 64289 Darmstadt, DE

[2] Institut für Elektromechanische Konstruktionen, Technische Universität Darmstadt, Merckstr. 25, 64283 Darmstadt, DE

E-mail: torsten.abel@physik.tu-darmstadt.de

Abstract. In the field of laser driven particle acceleration, many experiments with plain foil targets have been performed to investigate ion acceleration mechanisms such as TNSA and BOA. A more complex target geometry, for which the front and rear surface can be structured in a deterministic way, will lead to a better understanding of e.g. the actual source size, laser contrast related effect to the coupling efficiency of laser energy into the target of specifically shaped target surfaces or means to reduce the particle beam divergence such as ballistic focusing. A promising technology to manufacture complex 3D freeform targets is the two-photon-polymerization. With resolutions better than 200 nm, it is possible to create small targets with microstructured surfaces. We implemented the combination of two-photon-polymerization and UV-lithography to facilitate the handling of such small structures.

1. Introduction

So far, in the field of laser driven ion acceleration, a large variety of experiments have been performed with flat foil targets to investigate the fundamental processes involved [1, 2, 3, 4]. To expand the range of applications for the laser-generated particle beams (e.g. proton therapy [5], x-ray Thomson scattering [6]), the interest in manufacturing more complex, freeform targets increases. Simulations show that the highest ion energies are achieved with target sizes that are in the order of few times the laser spot size [7, 3]. It should be noted that targets in the size equal to the laser focus diameter are difficult to strike due to the pointing stability of the laser system. Given the fact that the diameter of the laserspot can be as small as several micrometers, it is challenging to manufacture freeform targets of similar size by using classic micro machining processes.

But not only the total size is of interest. We aim for small targets with a substructure like the sinusoidal pattern on the concave side of the hemispherical shape shown in figure 1. Flat foil targets with such a structure can be used to determine interesting parameters like ion sources size and divergence. In the same way targets like the conceptual design in figure 1 could be used to explore these parameters in combination with focusing effects. The outer diameter should be in the order of the incident lasers focus size. The whole target should be bigger than one times the laser focus due to typical pointing stabilities of current high power laser systems. The wavelength of the sinusoidal microstructure should be sufficiently small in order to observe

IOP Conf. Series: Journal of Physics: Conf. Series **1079** (2018) 012012 doi:10.1088/1742-6596/1079/1/012012

several periods across the source size. However, those targets can be created by two-photon-polymerization. The idea using two-photon-absorption to manufacture arbitrary 3D geometries dates back to the 90s [8]. Since that time it was used for many applications in different fields of research [9, 10].

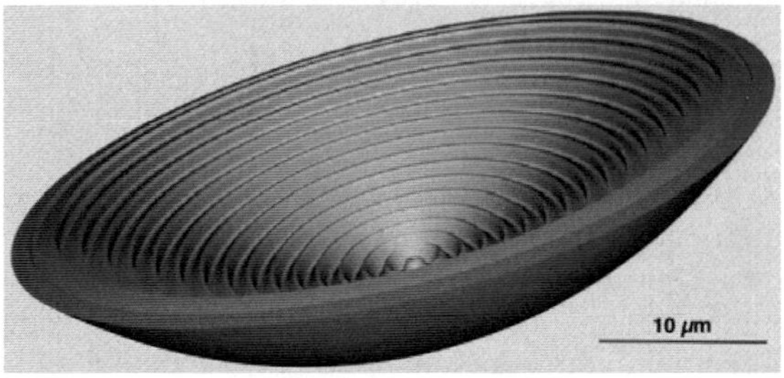

Figure 1. Conceptual design of an hemisphere target with inner sinusoidal microstructure. The outer diameter should be in the order of several times the incident lasers focus size. Due to the pointing stability of the laser the whole target should be bigger than one times the laser focus. The wavelength of the sinusoidal microstructure has to be smaller than the laser spot size.

The bare target can not be handled because of the small size. To attach the small targets to different standardized holders, as used in the respective experimental facilities, some kind of handle or frame is needed. Due to the fact that the expenditure of time for manufacturing strongly correlates with the target volume, the two-photon-polymerization is not suitable for large geometries. We managed to combine classic UV-lithography with the two-photon-polymerization, to shorten the manufacturing time.

2. Experimental Setup

With the two-photon-polymerization it it possible to manufacture small 3D freeform targets in the range of some μm [11]. But without some kind of holder it is nearly impossible to handle such small structures. It is too time-consuming to expose holding structures in the scale of some mm^3 with the two-photon-polymerization, so we use UV-lithography to create the comparatively large holding structures.

2.1. Substrate preparation

One of the most common substrates used with photoresist is Si-wafer, but in principle all kinds of flat homogeneous substrates are feasible. As we need a sacrificial layer to lift the exposed structures from the substrate, we used two distinct substrates with different metallic sacrificial layers. The first combination was a *FR-4* substrate with 35 μm copper on top, the second 18 μm aluminium on glass. *FR-4* is a glass-reinforced epoxy laminate. The substrate with copper on top is a standard source material for printed circuit boards and commercially available. Aluminium works well as sacrificial layer in combination with the *AZ 125 nXT* photo resist because it is solvable in the developer (*AZ 726MIF*) we used for this resist.

We use the same photoresist for the two-photon-polymerization and the UV-lithography. Most of the resists are designed to be used with an extensive UV light source. Traditionally, mercury vapour lamps have been used in lithography and many available photoresists are optimized for wavelengths at the $g-$, $h-$ and $i-$spectral lines from Hg. The g-line has a wavelength of 435 nm, the h-line is 404 nm and the i-line is 365 nm. These wavelengths fit well to a frequency doubled Titanium:Sapphire laser. We used two different types of photoresists

IOP Conf. Series: Journal of Physics: Conf. Series **1079** (2018) 012012 doi:10.1088/1742-6596/1079/1/012012

with different characteristics (*AZ 125nXT* and *SU-8 2025*). The *SU-8 2015* resist is an epoxy based resist with a proportion of solids of 68.55 % and a density of 1.219 g/ml. The ingredients of the AZ resist are not known. We choose negative resists to minimise the volume to be exposed and the exposure time. With this approach the thickness of the photo resist limits the target extension in one dimension. With both resists we could easily achieve film thicknesses thicker than 100 µm. To prepare the resist for the exposure regardless of which process, UV-lithography or two-photon-polymerization, a plane, thin layer of resist spread on a substrate is needed. We used spin coating to obtain such a layer. Spincoating means to spread the resist by rotating the substrate. The combination of the acceleration, the speed of rotation and the viscosity of the resist determines the resist thickness. To achieve a 100 µm thick layer of *AZ 125nXT* resist on a 20 mm × 20 mm rectangular glass substrate, we used the routine shown in table 1.

step	revolution per s	duration in s
1	300	10,0
2	1000	1,2
3	800	12,0
4	2000	1,5

Table 1. Programm steps to spincoat 100 um *AZ 125nXT* photo resist on a 20 mm rectangular glass substrate.

2.2. Exposure
The standard setup for UV-lithography consists of the photoresist, a mask and the UV-lightsource. The mask covers all the parts of the resist which should not be exposed. In the case of a negative resist, like in our setup, that means the covered parts will be removed by the development process after the exposure. With that method it is possible to fabricate so called 2.5D structures. 2.5D means the 2D pattern from the mask is extruded in the third dimension. The whole wafer area will be exposed in one step, thereby allowing the creation of multiple target frames in one step.

Instead of an extensive mask and UV-light source, which are used in conventional lithography, a focused laser with about 800 nm center wavelength comes into operation for the two-photon-polymerization. To expose the resist a nonlinear optical process, the two-photon-absorption is used. In lieu of one photon with 400 nm, two photons with 800 nm are absorbed, which results in the same energy transfer, as can be seen in Fig. 2.

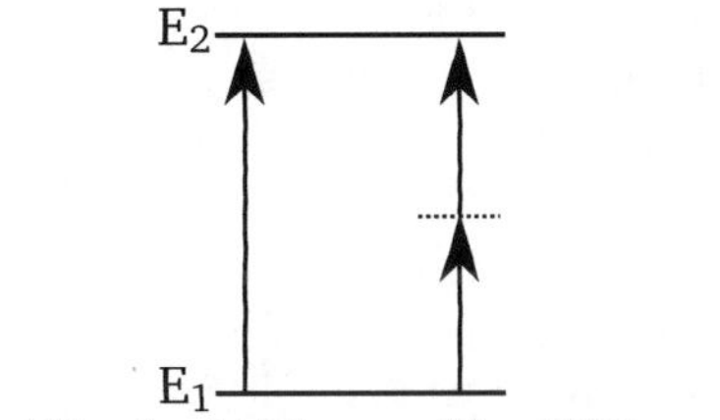

Figure 2. The simultaneous absorption of two photons with 800 nm is equal to the same energy transfer with the absorption of one photon at 400 nm.

For a simultaneous absorption of two photons, a high photon density is needed because of the quadratic dependence of two-photon absorption on the intensity of the incident laser beam [12]. Two-photon-absorption is related to $\chi^{(3)}_{\mathrm{imag}}$, the imaginary part of the third order susceptibility which is a material depended nonlinear optical property. Simultaneous absorption means in

IOP Conf. Series: Journal of Physics: Conf. Series **1079** (2018) 012012 doi:10.1088/1742-6596/1079/1/012012

this case, that the first absorbed photon creates a virtual state and the second photon must be absorbed during the lifetime of this state, which is approximately 10^{-15} s [13]. Therefore, this phenomenon only takes place when the light intensity is high enough. This condition can be found only in a small volume around the point of maximum intensity within the focus region, both laterally and along the direction of laser propagation. This small volumen is marked in figure 3.

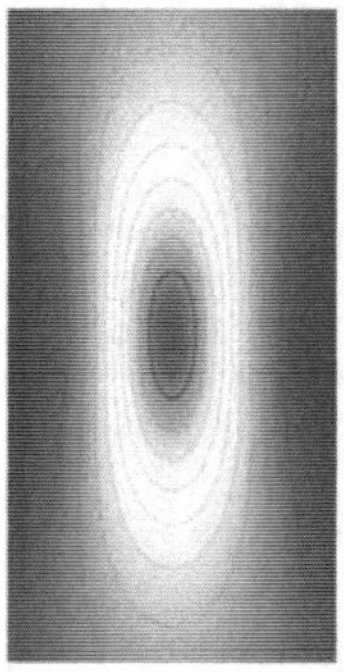

Figure 3. Image of an focused beam with iso-intensity lines. The solid contour marks the voxel.

Such a 3D volumetric pixel in which the exposure gets triggered is named a *voxel*. To create a 3D freeform by two-photon-polymerization, the resist has to be exposed point by point, respectively voxel by voxel. By increasing the voxel size, which leads to a reduction of the resolution, it is to a certain extend possible to reduce the process time.

In our setup we move the substrate with three high precision linear stages. Within one layer, we achieve 80 nm bidirectional accuracy over a travel range of 160 mm in X-direction and 50 mm in Y-direction. To reach a small voxel size, we use an objective with a numerical aperture of 0.85 and a working distance of 1.1 mm to focus the Ti:Sapphire laser into the resist. With our setup, we aim to achieve a radial resolution in the order of 200 nm. Along the laser propagation, the extension of the exposed volume pixel is typically larger, as compared to the radial dimension, and depends e.g. on the focusing geometry, exposure time, and laser intensity [14].

3. Multi component target manufacturing process

The first step of creating a target in a frame, made from photoresist, is to create a CAD model of both structures. In the following, two different frame designs for the same hemi-cone target are presented. We manufactured two different designs, using different resists depending on ongoing applications based on the final structures. The further application will be discussed elsewhere. The shown hemi-cone structure was chosen for demonstration purposes. To this end the structure in the frame is rotated by 90° and not hollow. The current orientation, as depicted in figure 4, eases the successive characterization with respect to the measurement of relevant geometrical features.

The target-frame combinations are designed to be exposed with the two previously described techniques and different resolutions. In figure 4, an example design for a fractionally manufactured target in a frame is shown. The hemi-cone target in the middle (red) and the two retaining bars (yellow), which connect the target and the ring in the frame, were manufactured with two-photon-polymerization. The frame (green), which is designed as a 2.5D object, was exposed by conventional UV-lithography.

The dimensions of the frame are displayed in figure 5. The frame was manufactured out of a 100 µm thick layer of *AZ125nXT*. The resist was spincoated on a glass substrate and prebaked for 40 min at 140 °C. A 13 µm aluminium layer between glass and resist worked as sacrificial layer.

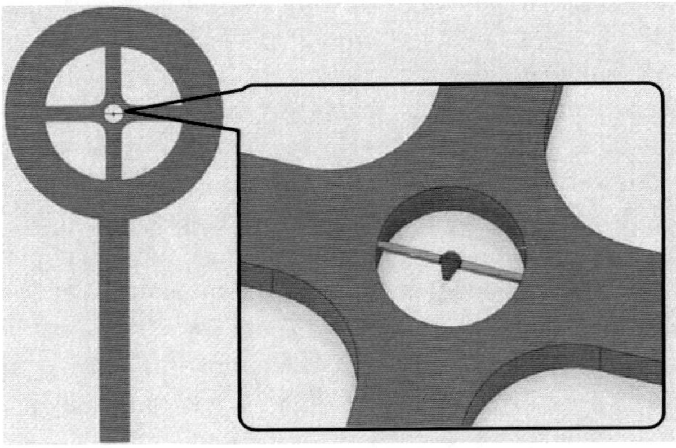

Figure 4. CAD model of a multi part target in a handling frame. The target (red) and the retaining bars (yellow) are exposed via two-photon-polymerization with different resolutions to be time efficient. The frame (green) is 2.5D and should be manufactured with UV-lithography.

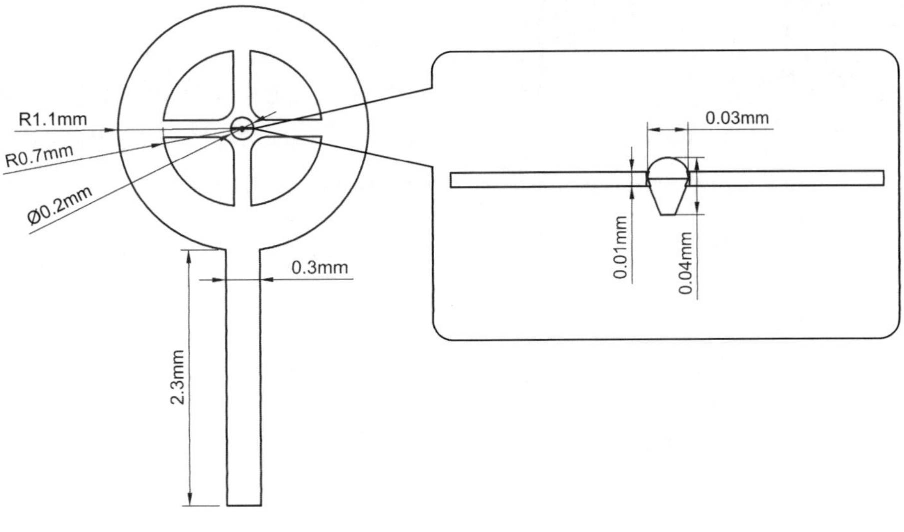

Figure 5. Dimensions of the designed target and its frame as shown in figure 4.

The sample was exposed with a light dose of $5.76\,\mathrm{J/cm^2}$. As the mask-aligner we used for the exposure does not have any filters available, we used the *g*-,*h*-and *i*-line of the Hg-spectrum.

For the two-photon-polymerization we used a Titanium:sapphire laser with a central wavelength of 800 nm, a repetition rate of 80 MHz and typically a pulswidth of 100 fs. Depending on the resolution and processing time, we can provide an average laser power in the range of 5 mW to 50 mW with a movement speed of $0.1\,\mathrm{mm\,s^{-1}}$ to $10\,\mathrm{mm\,s^{-1}}$.

Since the target shown in red (figure 4) is the most important part, a high line resolution

IOP Conf. Series: Journal of Physics: Conf. Series **1079** (2018) 012012 doi:10.1088/1742-6596/1079/1/012012

of 200 nm was used. This high geometrical resolution was achieved by a movement speed of 0.3 mm s^{-1} and 20 mW laser power. The combination of low speed and high number of lines results in a very time consuming process. For the retaining bars, the resolution is of less interest, since they just have to connect the target to the frame. The aim of the design must be to reduce the diameter of these bars to find the thinnest bars which withstand the forces during development, to further reduce the process time.

Once the liquid is dried and the bars do only need to support the target itself, structural integrity is not an issue any more, as the weight of the target does not impose any significant force on the interconnection bars. Right now, the bars are 10 µm to make sure they are strong enough. Future designs will further reduce this part. The bars were exposed with a line resolution of 500 nm and a movement speed of 1 mm s^{-1} and a laser power of 50 mW to speed up the process.

The result of the combined exposure with two-photon-polymerization and UV-lithography is shown in figure 6.

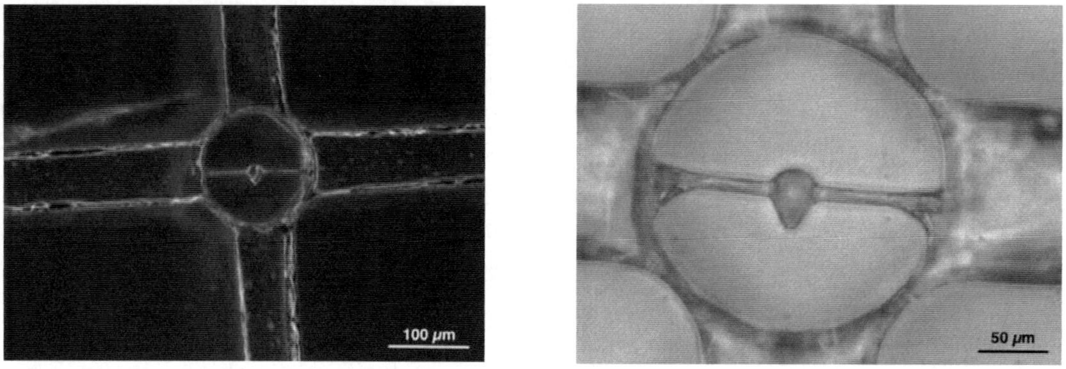

Figure 6. Manufactured target, following the design shown in figure 4. The hemi-cone target and the retaining bars were created with two-photon-polymerization with the target structure having a higher resolution. The handling frame was manufactured with conventional UV-lithography.

For a second application, we prepared 80 µm of *SU-8* on a *FR-4* substrate with 36 µm copper on top. To make sure that the UV exposed structure overlaps with the target stilts, alignment marks were needed, which were created by partially removing the sacrifical layer. The mark consists of a circle in the middle of a cross as can be seen in figure 7. The circle has a diameter of 250 µm.

After spincoating, a first prebake of 3 min at 105 °C was done, before the target was build via two-photon-polymerization. After the two-photon-polymerization, a second bake of 180 min at 105 °C minimized the solvent content. In contrast to the exposure of the first frame design, here the large holding structure was exposed with 800 mJ/cm^2 light with only one wavelength of 365 nm corresponding to the Hg-i-line. After the exposure, a postbake at 95 °C for 60 min followed. We used Propylene glycol methyl ether acetate (PGMEA) for 10 minutes to develop the structure. After the development process, some resist was left around the target. To remove all the remains, the samples were plasma etched. The etching unit has a power of 1000 W and was used for 5 min at 25 °C. The process gases were 1000 sccm of O_2, 50 sccm of CF_4 and 50 sccm of N_2.

4. Conclusion

We successfully combined the two-photon-polymerization with the UV-lithography in the field of targetry. With the combination of these techniques, we made it possible to produce and

Figure 7. The target frame, made of SU-8 photo resist, has a hole of 250 µm in the middle to enable a free-standing target. In the background, the Cu sacrificial layer is visible. The cross visible through the photoresist serves as an alignment mark. It was prepared by removing the sacrificial layer.

handle micro structured targets in the size of only a few micrometers, safely held by a larger supporting structure. With the two-photon-polymerization exposure and successive processing, one can fabricate structures in true 3D with sufficiently high resolution to be used for many applications in the realm of targetry. With UV-lithography, it is possible to expose target handling frames in the size of several mm^2 within minutes. Using both lithographic processes for one target is a promising method for high number production of complex micro targets as this approach combines the batch processing at medium resolution during conventional mask exposure lithography with the ultra high resolution and true free form exposure with the two-photon-polymerization.

5. Acknowledgement

This work was financially supported by the DFG in the framework of the Excellence Initiative, Darmstadt Graduate School of Excellence Energy Science and Engineering (GSC 1070).

References

[1] Snavely R A *et al.* 2000 *Physical Review Letters* **85** 2945–2948 ISSN 00319007
[2] Hatchett S P *et al.* 2000 *Physics of Plasmas* **7** 2076 ISSN 1070664X
[3] Schollmeier M S 2008 *Optimization and control of laser-accelerated proton beams* Dr. rer. nat. TU Darmstadt
[4] Wagner F *et al.* 2016 *Physical Review Letters* **116** 1–5 ISSN 10797114
[5] Bulanov S V and Khoroshkov V S 2002 *Plasma Physics Reports* **28** 453–456 ISSN 1063-780X
[6] Glenzer S H and Redmer R 2009 *Reviews of Modern Physics* **81** 1625–1663 ISSN 15390756
[7] Daido H, Nishiuchi M and Pirozhkov A S 2012 *Reports on Progress in Physics* **75** 72 ISSN 0034-4885
[8] Maruo S, Nakamura O and Kawata S 1997 *Optics letters* **22** 132–4 ISSN 0146-9592
[9] Koroleva A, Gill A A, Ortega I, Haycock J W, Schlie S, Gittard S D, Chichkov B N and Claeyssens F 2012 *Biofabrication* **4** 025005 ISSN 1758-5082
[10] Schlie S, Ngezahayo A, Ovsianikov A, Fabian T, Kolb H A, Haferkamp H and Chichkov B N 2007 *Journal of Biomaterials Applications* **22** 275–287 ISSN 0885-3282
[11] Maruo S and Kawata S 1998 *Journal of Microelectromechanical Systems* **7** 411–415 ISSN 10577157
[12] Lee K S, Kim R H, Yang D Y and Park S H 2008 *Progress in Polymer Science (Oxford)* **33** 631–681 ISSN 00796700
[13] Sanna Turunen (Tampere University of Technology) 2017 Direct Laser Writing of Proteins and Synthetic Photoresists for Neuronal Cell Growth Guidance Phd Tampere University of Technology
[14] Lee K s, Yang D y, Park S H and Kim R H 2006 *Wiley InterScience* 72–82

IOP Conf. Series: Journal of Physics: Conf. Series **1079** (2018) 012013 doi:10.1088/1742-6596/1079/1/012013

Automatic rough alignment for key components in laser driven experiments using fiducial markers

D C Popescu[1,2], M O Cernaianu[1] and I Dumitrache[2]

[1] Extreme Light Infrastructure - Nuclear Physics (ELI-NP) / Horia Hulubei National Institute of Physics and Nuclear Engineering - Reactorului St., 30, PO Box MG-6, 077125, Bucharest-Magurele, Romania

[2] Faculty of Automatic Control and Computers, University Politehnica of Bucharest, 313 Spl. Independentei, 060042 Bucharest, Romania

E-mail: dragos.popescu@eli-np.ro / @acse.pub.ro

Abstract.
In the laser - solid target experiments at ELI-NP, maximizing the availability of the laser system and hence the number of shots during a campaign is envisaged. One of the factors that affects this is the time for preparing the experiment and the time for changing the targets between shots. In this paper we present a method for automatic rough alignment of various experimental setup instruments like the target frames, microscopes and diagnostics. Giving the impossibility to define a global reference system that captures precisely enough the relative position of all equipment in a continuously changing experimental setup, the method developed is using relative optical measurements using fiducial markers. The method is conceived to be used along with other micron-level alignment methods as a pre-alignment phase. Along this work, a setup and the algorithms were developed in order to test the accuracy of the method. First tests revealed a rough alignment accuracy between 100-200μm which was achieved using only a low resolution imaging camera. The accuracy can be further improved using higher resolution imaging systems or by developing new key points in the algorithm that are discussed at the end of the paper.

1. Introduction

For laser - solid target experiments where the accelerated particle beams require reproducibility and enhancement of the parameters (e.g. maximum energy, spatial homogeneity, angular distribution, energy spectrum), preparing and aligning the experimental setup, including the target is a highly demanding task. In general, running the experiments requires remote control as the interaction takes place in vacuum, while the generated ionizing radiation requires the shielding in heavy concrete bunkers [1].

Considering the high-repetition rate of current laser systems (on the order of up to tens of shots per second), the availability of the targets have to be also ensured. This can be affected by long experiment preparation times and by the fact that for changing the target it is required to vent and pump down the interaction chamber. For the first issue, when solid targets are used, the only solution is to implement fast alignment and target exchanging methods. The latter issue can be solved by using a target feeding system which stores multiple target frames under vacuum [2], [3].

The peak intensity of the laser on the target is a crucial parameter that can be affected by the deviations of the target position. For the case where tight focusing is used to generate high-intensity optical fields, the Rayleigh length of the focused laser beam is in the range of micrometers to tens of micrometers. So, in addition to fast alignment, very precise alignment is required.

At the ELI-NP facility, different experiments with multi PW laser beams will be operated [4], [5] with various equipment in the interaction chambers. The continuously changing setups lead to time consuming operation if manual pre-alignment and calibration procedures are used where physical absolute references (like metallic stalks or optical irises) have to be installed.

In previous works, absolute reference alignment methods have been developed achieving good results. The retro imaging system [6], [7], [8] uses the reflection from the target which is imaged back through the beam transport. The image produced by the target in the right position is used as reference image for positioning other targets.

The rear surface imaging method [9], [10], [11] uses a microscope for imaging the focal spot. Its focal plane becomes the reference for positioning the target.

The method was improved with the addition of a near-field and a far-field imaging for increasing the precision of orientation alignment [12]. For speeding up the alignment procedure, the microscope was improved by adding one extra optical systems with a lower magnification [13]. For increasing the accuracy, the microscope was coupled with multiple interferometers [14] achieving micron-level alignment.

Another solution was the use of a setup with three orthogonal lasers which create through three orthogonal holes Fresnel Diffraction images used as references for positioning new targets [15]. All these methods either use setups with fixed and well defined structure or need complex pre-alignment and calibration procedures which are time consuming.

In this work we introduce an automatic rough alignment method and positioning algorithm [16] based on a numerical real-time optimization procedure which aims to reduce the time necessary for preparing the setup of an experiment. The method is based on optical position measurements for fiducial markers which are used for building relative coordinate systems. This method is suitable for automatic pre-alignment of instruments with motorized stages, however, it can provide support also for manual alignment.

2. Automatic alignment system concept

The fiducial markers used in Augmented Reality applications hold an unique ID, a number between 1 and n (depending on how large the dictionary used is), which can be detected and retrieved from images using a processing algorithm (in a similar manner with bar-codes or QR-codes). In addition, their position and orientation with respect to the camera used for imaging can be estimated. This is achieved with the aid of the projective geometry which provides a pinhole camera model that is expressing the relationship between physical 3D points and their 2D projections from images [17].

For implementing the alignment method, ARUCO markers were chosen, given a complete software library with tools for generating and detecting the markers is available [18], [19]. In figure 1 is shown an ARUCO fiducial marker where the ID is detected by the library, while the position and orientation is estimated.

It is considered that the camera holds a virtual coordinate system having the z axis normal to the image sensor ($\vec{x_c}\vec{y_c}\vec{z_c}$ - figure 1). The fiducial marker has a similar coordinate system attached ($\vec{x_m}\vec{y_m}\vec{z_m}$). The ARUCO algorithm is estimating the position and the orientation of the marker coordinate system relative to the camera coordinate system. This is given by a homogeneous transformation denoted as T_c^m (the transformation between the camera coordinate system and the marker coordinate system) which is defined as:

$$T_c^m = \begin{bmatrix} r_{11} & r_{12} & r_{13} & x \\ r_{21} & r_{22} & r_{23} & y \\ r_{31} & r_{32} & r_{33} & z \\ 0 & 0 & 0 & 1 \end{bmatrix} \tag{1}$$

where $R = \begin{bmatrix} r_{11} & r_{12} & r_{13} \\ r_{21} & r_{22} & r_{23} \\ r_{31} & r_{32} & r_{33} \end{bmatrix}$ is a rotation matrix which expresses the orientation of the $\vec{x_m}\vec{y_m}\vec{z_m}$ coordinate system relative to the orientation of the $\vec{x_c}\vec{y_c}\vec{z_c}$ coordinate system while x, y, z expresses their relative position.

Considering that $\vec{x_c}, \vec{y_c}, \vec{z_c}, \vec{x_m}, \vec{y_m}, \vec{z_m}$ are versors (orthogonal unit vectors) it can be written that:

$$\begin{cases} r_{11} = \vec{x_m} \cdot \vec{x_c} \\ r_{12} = \vec{y_m} \cdot \vec{x_c} \\ r_{13} = \vec{z_m} \cdot \vec{x_c} \\ r_{21} = \vec{x_m} \cdot \vec{y_c} \\ r_{22} = \vec{y_m} \cdot \vec{y_c} \\ r_{23} = \vec{z_m} \cdot \vec{y_c} \\ r_{31} = \vec{x_m} \cdot \vec{z_c} \\ r_{32} = \vec{y_m} \cdot \vec{z_c} \\ r_{33} = \vec{z_m} \cdot \vec{z_c} \end{cases} \tag{2}$$

where the operator "·" denotes the dot product. Also, the two origins (O_c and O_m) are defining the following position vector:

$$O_c\vec{O}_m = x \cdot \vec{x_c} + y \cdot \vec{y_c} + z \cdot \vec{z_c} \tag{3}$$

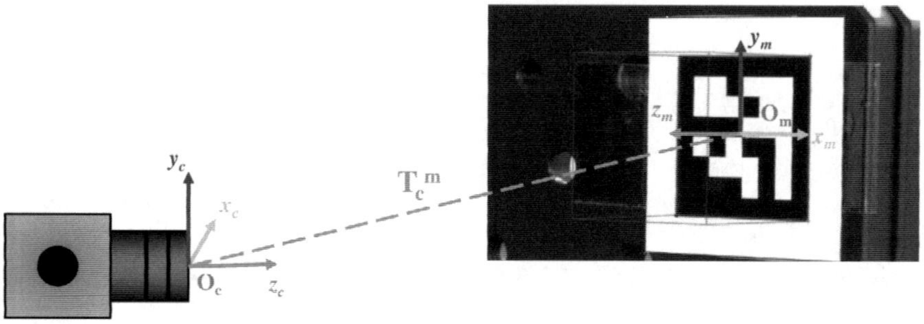

Figure 1. The estimation of the position and orientation of an ARUCO [18], [19] fiducial marker and ID identification using an imaging camera. This is achieved using the pinhole camera model in which the extrinsic parameters are determined knowing the intrinsic ones [17].

In general, the experimental setups are positioned using absolute references which rely on precise features in the interaction chamber (ex. the target chamber center, metallic stalks or optical irises). In the case of solid targets, the positioning of the targetry equipment relative to another, like the alignment microscope, would be desired. In this approach of relative positioning, the target can be automatically pre-aligned without requiring to align the target

IOP Conf. Series: Journal of Physics: Conf. Series **1079** (2018) 012013 doi:10.1088/1742-6596/1079/1/012013

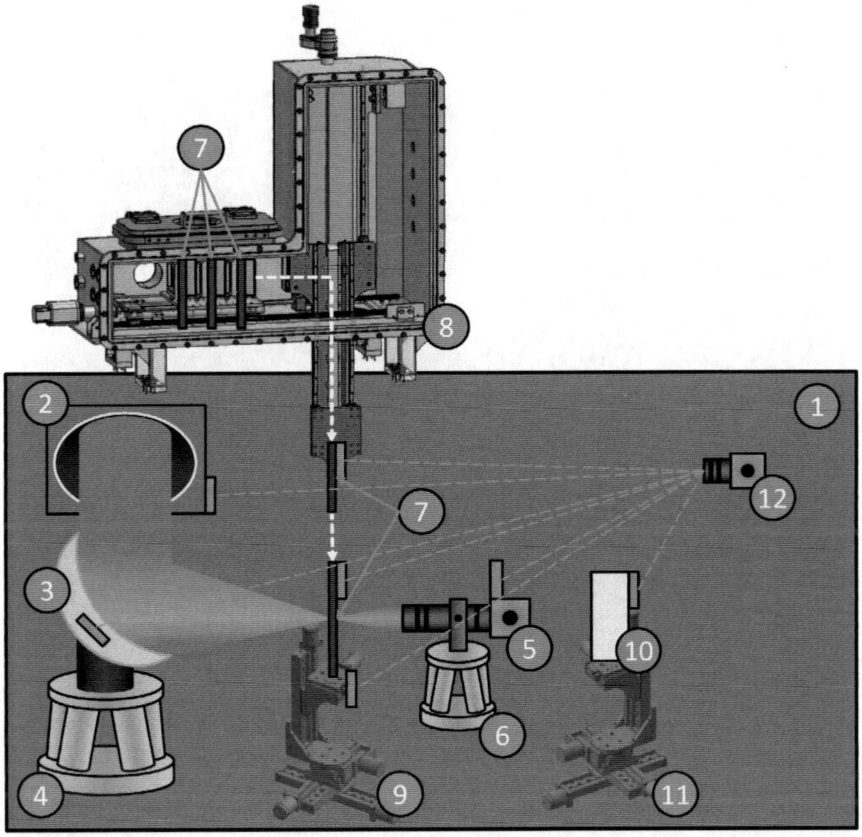

Figure 2. Experimental setup example and pre-alignment system using fiducial markers. The experimental setup is composed by: interaction chamber (1), laser beam input flange (2), off-axis parabolic mirror (OAP) (3), OAP manipulator (4), optical microscope (5), microscope manipulator (6), target frames (7), target feeding system (8), target frame manipulator (9), detector (10), detector manipulator (11), imaging camera (12). The fiducial markers attached on each relevant instrument are drawn in blue.

to the laser beam. The laser focal spot is optimized and its position is determined using an alignment microscope which becomes the reference position for pre-aligning the target. This method allows also to automatically exchange the target from the manipulator using a target feeding system whose position and configuration was changed without the need of a calibration procedure. Moreover, it is possible to automatically position instruments with motorized stages in positions relative to the target frame or relative to another instrument.

In figure 2 is presented an example of experimental setup where the instruments are pre-aligned relative one to each other's positions. The proposed method is used for rough alignment and coupled with the rear surface imaging method for fine alignment. It is considered that in the interaction chamber (1) all relevant instruments have attached on them a fiducial marker (drawn in blue), which are used to build relative reference positions. The experimental setup is composed by: laser beam input flange (2), off-axis parabolic mirror (OAP) (3), OAP manipulator (4), optical microscope (5), microscope manipulator (6), target frames (7), target feeding system (8), target frame manipulator (9), generic detector (10), detector manipulator (11), imaging camera (12). The camera used for imaging needs to be fixed before each instrument pre-alignment, but

IOP Conf. Series: Journal of Physics: Conf. Series **1079** (2018) 012013 doi:10.1088/1742-6596/1079/1/012013

it can be moved conveniently from one instrument to another especially when working in large interaction chambers.

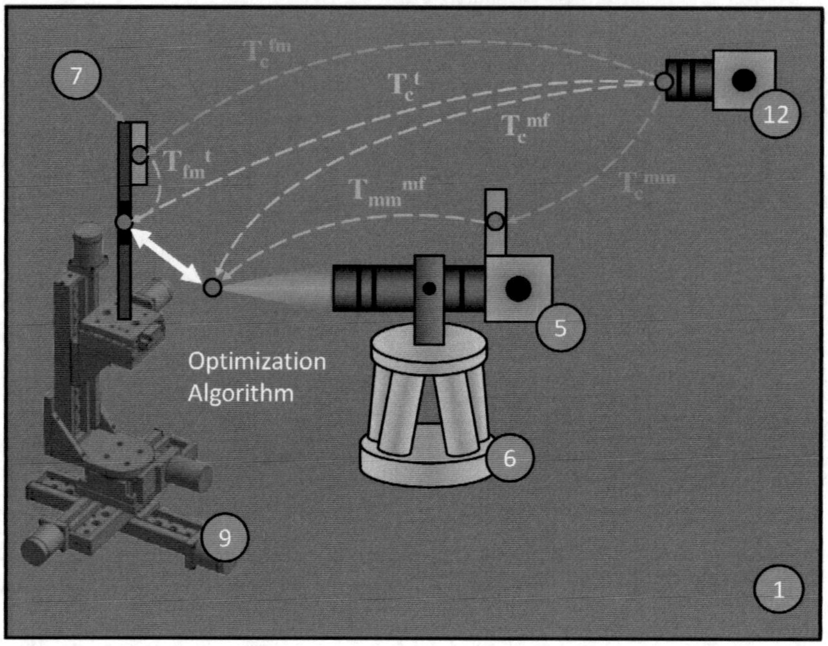

Figure 3. Target pre-alignment in microscope's field of view. The figure illustrates how the relative reference positions and virtual coordinate systems can be defined. The virtual coordinate systems considered are illustrated simplified through red circles accompanied by a set of letters for reference (c - the virtual coordinate system attached to the camera; mm - attached to the fiducial marker on the microscope; mf - attached to the center of the focal plane of the microscope; fm - attached to the fiducial marker on the frame; t - attached to the target). Between these coordinate systems are defined homogeneous transformations which partially are known a priori, and partially are determined in real time using the fiducial markers.

As a clarifying example, in figure 3 is presented an approach for considering and defining the virtual coordinate systems and the relative reference positions for pre-aligning the target in microscope's field of view (FOV). This choice is not unique, the user has to build his own setup while deciding how to consider the virtual coordinate systems and where to place the fiducial markers, developing calibration procedures accordingly. The instruments are marked using the same numbers as in figure 2 while the markers are also drawn in blue. The virtual coordinate systems considered are illustrated simplified through red circles accompanied by a set of letters for reference (c - the virtual coordinate system attached to the camera; mm - attached to the fiducial marker on the microscope; mf - attached to the center of the focal plane of the microscope; fm - attached to the fiducial marker on the frame; t - attached to the target).

T_c^{mm} defines the relative position and orientation between the camera coordinate system and the coordinate system attached to the fiducial marker on the microscope. The center of the focal plane of the microscope (MCFP) has a well defined position determined by the microscope's construction (sensor and optical system characteristics). T_{mm}^{mf} defines this relative position and orientation between the coordinate system attached to the fiducial marker on the microscope and the coordinate system attached to the center of the focal plane. This homogeneous transformation will remain unchanged as long as the microscope assembly is not

IOP Conf. Series: Journal of Physics: Conf. Series **1079** (2018) 012013 doi:10.1088/1742-6596/1079/1/012013

Table 1. The virtual coordinate systems and homogeneous transformations for target pre-alignent in microscope's FOV.

Virtual coordinate system A	Virtual coordinate system B	A to B transformation	Source
Camera	Microscope marker	T_c^{mm}	Using ARUCO library
Microscope marker	MCFP	T_{mm}^{mf}	Manufacturing or calibration
Camera	MCFP	$T_c^{mf} = T_c^{mm} \cdot T_{mm}^{mf}$	Computed in real-time
Camera	Frame marker	T_c^{fm}	Using ARUCO library
Frame marker	Target	T_{fm}^{t}	Manufacturing or calibration
Camera	Target	$T_c^{t} = T_c^{fm} \cdot T_{fm}^{t}$	Computed in real-time

modified. It can be obtained either through precise design and manufacturing of the microscope, either through a calibration procedure using a dedicated calibration setup. Consequently, the relative position and orientation between the camera coordinate system and the coordinate system attached to the center of the focal plane is defined as:

$$T_c^{mf} = T_c^{mm} * T_{mm}^{mf} \tag{4}$$

For the target frame, T_c^{fm} defines the relative position and orientation between the camera coordinate system and the coordinate system attached to the fiducial marker on the frame. T_{fm}^{t} defines the relative position and orientation between the coordinate system attached to the fiducial marker on the frame and the coordinate system attached to the first target from the frame. This homogeneous transformation is determined by the frame construction and where the fiducial marker is placed and will remain unchanged as long as the frame or the target wafer structure is not altered. Consequently, the relative position and orientation between the camera coordinate system and the coordinate system attached to the first target from the frame is defined as:

$$T_c^{t} = T_c^{fm} * T_{fm}^{t} \tag{5}$$

The manner in which the coordinate systems and the homogeneous transformations are considered and defined is summarized in table 1.

The target is pre-aligned in the microscope's FOV when the two homogeneous transformations T_c^{mf} and T_c^{t} are made equal by moving the motorized manipulators.

In the same manner, the OAP can be automatically pre-aligned whenever the focusing configuration is changed, in relative positions to the laser beam input flange.

For exchanging the target using the feeding system, the target manipulator needs to be placed in a precise position in order to ensure the interface between the two motion systems. The manipulator can be pre-aligned using two fiducial markers attached on the two instruments.

Typically, any type of detector that is moved using a motorized manipulator can be pre-aligned in a fixed position relative to the interaction point by using a fiducial marker attached on it and the one from the target frame.

3. Automatic alignment system prototype

The relative reference positions are build using fiducial markers attached on targets and instruments. In order to achieve the pre-alignment as described, by imposing the equality of two

IOP Conf. Series: Journal of Physics: Conf. Series **1079** (2018) 012013 doi:10.1088/1742-6596/1079/1/012013

homogeneous transformations, we developed an algorithm that iteratively searches the positions along all the stages from a manipulator until the equality condition is met. The algorithm runs in real-time and at its core it solves an optimization problem using the numerical gradient descent method [20] where the gradient is estimated in real time at each iteration [21].

In a previous work [22] the accuracy of measuring the position and orientation of an ARUCO fiducial marker was experimentally determined. By using a 140 μm optical resolution imaging setup, it was obtained a measurement accuracy along the X and Y axis of approximately 75 μm. Along the Z axis, the accuracy was about 300 μm.

In order to test the accuracy of the pre-alignment method and the positioning algorithm, we built a test setup, as sketched in figure 4. The core is an embedded computing module (Nvidia Jetson TX1) which runs the ARUCO library and the real-time positioning algorithm.

The manipulator has 5 degrees of freedom (3 translations, one rotation and one goniometer) and is assembled from vacuum compatible Standa stages having a resolution of 2.5 μm/step and 0.6 $arcmin$/step respectivelly. It is controlled using Standa controllers and a computer running a Tango server which receives commands from a Tango client running on the Jetson device. On top of the manipulator an ARUCO marker is attached. For imaging, a 2 Mpix camera with a 25 mm lens was used. The distance between the camera and the fiducial marker is around 80 cm.

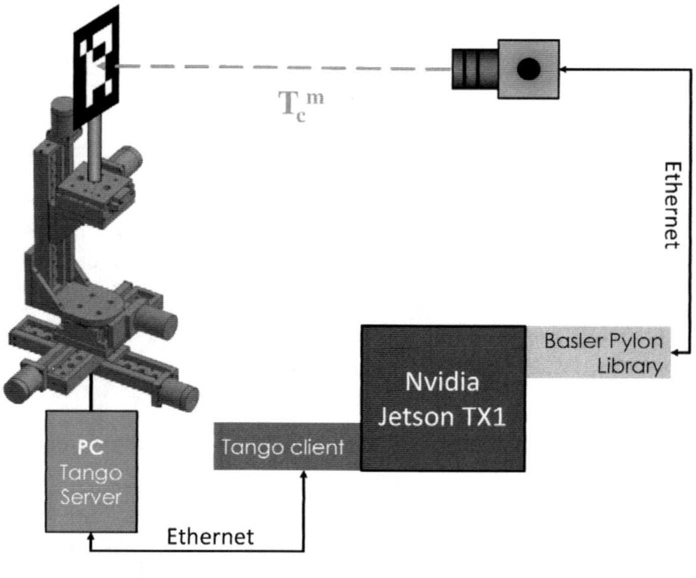

Figure 4. Pre-alignment system prototype setup. The core of the prototype is an embedded computing module, Nvidia Jetson TX1. It runs the ARUCO library for estimating the position and orientation of the fiducial marker, the Tango client that interacts with the motorized manipulator, the Basler Pylon library for controlling and acquiring images from the camera and the real-time optimization algorithm for driving the positioning.

This test is considering only the translation positioning and not the orientation. The test procedure follows the next steps:

(i) the manipulator is moved to a fixed, known position ($X = 15000$, $Y = 20000$, $Z = 30000$ steps)

(ii) the position and orientation of the fiducial marker (T_c^m) is stored and considered the reference position

Table 2. Positioning accuracy results after tests. The quantities marked with * are given in motor steps.

Test no.	Position after 7 iterations X^*	Y^*	Z^*	Positioning error - manipulator ΔX^*	ΔY^*	ΔZ^*	$\Delta X[\mu m]$	$\Delta Y[\mu m]$	$\Delta Z[\mu m]$
1	14985	20008	30022	+15	-8	-22	+37.5	-20	-55
2	14993	19996	29965	+7	+4	+35	+17.5	+10	+87.5
3	14977	20000	29938	+23	0	+62	+57.5	0	+155
4	15018	20004	30070	-18	-4	-70	-45	-10	-175
5	15004	19998	29974	-4	+2	+26	-10	+5	+65

(iii) the manipulator is moved to its home position ($X = 0$, $Y = 0$, $Z = 0$ steps)

(iv) the real-time positioning algorithm is started. It needs to bring back the fiducial marker to the reference position without knowing that this corresponds to $X = 15000$, $Y = 20000$, $Z = 30000$ steps in manipulator coordinates

Five tests were made in order to prove the consistency of the results. For each test, the positioning algorithm ran for 7 iterations. The results are presented in table 2.

The results of the tests show that the positioning error along the X and Y axis is less than 60 μm, while along the Z axis is higher, at around 180 μm. This is consistent with the precision of measuring the position of the fiducial markers which along the Z axis is lower compared with the X and Y axis. Each positioning test lasts less than 4 minutes and after the first iteration the positioning error was less than 1 mm.

4. Conclusions

The pre-alignment method revealed during the first tests an accuracy between 100-200μm, achieved using only a 2 Mpix camera. Although the method was conceived to be used as an automatic pre-alignment phase in addition to another fine alignment method (ex. rear surface imaging), it can be further improved and used as stand-alone for applications where an accuracy of few tens of microns is enough.

The accuracy of the method is directly influenced by the accuracy in detecting the position and the orientation of a fiducial marker. A detailed study on this topic can be found in [22]. The accuracy scales linearly with the resolution of the imaging camera. Using, for example, a 10 Mpix camera in a similar setup can lead to a theoretical accuracy of up to 20μm. Furthermore, using different imaging camera optics can improve the results with the cost of reducing the working volume.

The optimization algorithm is also responsible for how accurate the method is. At the current stage of development, the optimization method based on gradient descent was implemented in a standard approach. Multiple improvements are envisaged for tuning in real-time the gradient which will increase the convergence, therefore, the alignment will be faster and more accurate.

Future works will focus also on extending the algorithm in using multi-camera setups in order to improve the accuracy along the Z axis and for expanding the working volume. For mitigating the effect of illumination on the accuracy, automatic camera calibration algorithms are also required to be developed.

Acknowledgments

This work is supported by Extreme Light Infrastructure Nuclear Physics (ELI-NP) - Phase II, a project co-financed by the European Union through the European Regional Development

Fund through the Comptitiveness Operational Programme Investing in Sustainable Development (1/07.07.2016, COP ID 1334). Part of the equipment was funded by the European Cluster of Advanced Laser Light Sources (EUCALL) project which has received funding from the European Unions Horizon 2020 research and innovation programme under agreement No. 654220.

References

[1] Mitu I O, Ivan C, Negoita F, Aranghel D, Bercea S, Iliescu E, Petcu C, Gugiu M, Popovici M A and Ur C A 2016 *Romanian Reports in Physics* **68** S885–S945
[2] Spindloe C, Tolley M K and Diva E 2009 *Central Laser Facility Annual Report*
[3] Green J S, Brummitt P A, Foster P S, Neville D R, Rajeev P P, Spindloe C, Streeter M J V, Tolley M, Neely D N, Alexander N B, Frey D, Salcedo J and Le G 2009 *Central Laser Facility Annual Report*
[4] Negoita F, Roth M, Thirolf P G, Tudisco S, Hannachi F, Moustaizis S, Pomerantz I, McKenna P, Fuchs J and Sphor K 2016 *Romanian Reports in Physics* **68** S37
[5] Turcu I C E, Murphy C, Negoita F, Stutman D, Zepf M, Schreiber J, Harvey C, Gray R J, Toma M and Balascuta S 2016 *Romanian Reports in Physics* **68** S145
[6] Carroll D C, Coury M, Scott G, McKenna P, Streeter M J V, Nakamura H, Najmudin Z, Fiorini F, Green S, Green J S, Foster P, Heathcote R, Poder K, Symes D, Clarke R J, Pattathil R and Neely D 2011 *Central Laser Facility Annual Report*
[7] Spencer I, Ledingham K W D, McKenna P, McCanny T, Singhal R P, Foster P S, Neely D, Langley A J, Divall E J, Hooker C J, Clarke R J, Norreys P A, Clark E L, Krushelnick K and Davies J R 2003 *Physical Review E* **67** URL `https://doi.org/10.1103%2Fphysreve.67.046402`
[8] McKenna P, Ledingham K W D, Spencer I, McCany T, Singhal R P, Ziener C, Foster P S, Divall E J, Hooker C J, Neely D, Langley A J, Clarke R J, Norreys P A, Krushelnick K and Clark E L 2002 *Review of Scientific Instruments* **73** 4176–4184 URL `https://doi.org/10.1063%2F1.1516855`
[9] Symes D, Booth N, Foster P, Anderson T, Marshall A, Rathbone D, Holligan P, Spurdle S, Neville D and Brumm P 2013 *Central Laser Facility Annual Report*
[10] Willis C, Poole P L, Akli K U, Schumacher D W and Freeman R R 2015 *Review of Scientific Instruments* **86** 053303 URL `https://doi.org/10.1063%2F1.4921554`
[11] Singh P K, Kakolee K, Jeong T and Ter-Avetisyan S 2016 *Nuclear Instruments and Methods in Physics Research Section A: Accelerators, Spectrometers, Detectors and Associated Equipment* **829** 363–366 URL `https://doi.org/10.1016%2Fj.nima.2016.01.057`
[12] Anderson T, Booth N, Foster P and Symes D 2013 *Central Laser Facility Annual Report*
[13] Symes D R, Booth N, Baraclough M, Indorf G, Oliver P, Scott G G, Neely D, Spindloe C, Heathcote R I and Clarke R J 2015 *Central Laser Facility Annual Report*
[14] Booth N, Clarke R, Heathcote R, Neely D, Pattathil R, Rusby D, Spindloe C, Symes D, Tolley M and Tomlinson S 2014 High-rep rate target development for ultra-intense interaction science at the central laser facility *Target Diagnostics Physics and Engineering for Inertial Confinement Fusion III* ed Bell P M and Grim G P (SPIE) URL `https://doi.org/10.1117%2F12.2061803`
[15] Alexander N B, Goodin D T and Stephens R B 2007 *Fusion Science and Technology* **51** 795–799 URL `https://doi.org/10.13182%2Ffst07-a1482`
[16] Popescu D C and Cernaianu M O 2017 Method and system for automatic positioning of elements for configurations of experiments with high-power lasers, Romanian patent pending no. RO132327 (A0)
[17] Hartley R and Zisserman A Camera geometry and single view geometry *Multiple View Geometry in Computer Vision* (Cambridge University Press) pp 151–152 URL `https://doi.org/10.1017%2Fcbo9780511811685.009`
[18] Garrido-Jurado S, Muñoz-Salinas R, Madrid-Cuevas F and Marín-Jiménez M 2014 *Pattern Recognition* **47** 2280–2292 URL `https://doi.org/10.1016%2Fj.patcog.2014.01.005`
[19] Garrido-Jurado S, Muñoz-Salinas R, Madrid-Cuevas F and Medina-Carnicer R 2016 *Pattern Recognition* **51** 481–491 URL `https://doi.org/10.1016%2Fj.patcog.2015.09.023`
[20] Boyd S and Vandenberghe L 2004 *Convex Optimization* (Cambridge University Press) URL `https://doi.org/10.1017%2Fcbo9780511804441`
[21] Srinivasan B, François G and Bonvin D 2011 Comparison of gradient estimation methods for real-time optimization *Computer Aided Chemical Engineering* (Elsevier) pp 607–611 URL `https://doi.org/10.1016%2Fb978-0-444-53711-9.50122-x`
[22] Popescu D C, Cernaianu M O, Ghenuche P and Dumitrache I 2017 An assessment on the accuracy of high precision 3d positioning using planar fiducial markers *2017 21st International Conference on System Theory, Control and Computing (ICSTCC)* (IEEE) URL `https://doi.org/10.1109%2Ficstcc.2017.8107079`

IOP Conf. Series: Journal of Physics: Conf. Series **1079** (2018) 012014 doi:10.1088/1742-6596/1079/1/012014

Experimental Testing of Targets for a High Accuracy Microtarget Supply (HAMS) System on the Gemini Laser System

C Spindloe[1,2], N Booth[1], S Astbury[1], C Gregory[1], E Bryce[1], S Tomlinson[1], R Heathcote[1], G Arthur[2], D Haddock[1], and M K Tolley[1,2]

[1]Science and Technology Facilities Council, Rutherford Appleton Laboratory, Harwell Campus, Chilton, Didcot, Oxon, OX11 0QX, UK
[2]Scitech Precision Ltd, Rutherford Appleton Laboratory, Harwell Campus, Chilton, Didcot, Oxon, OX11 0QX, UK

E-mail:christopher.spindloe@stfc.ac.uk

Abstract. It is widely understood within the high-power laser community that recent developments in diode pumped and high repetition rate laser systems will give unprecedented access to laser shots. This will provide a challenge for target fabrication in making enough experimental samples. While in the past access to facilities and shot rates during access periods have been the limiting factor for high power laser experiments this will soon not be the case. There has already been a shift in development of the user base from fundamental science experiments to industrial applications using the laser experiment as a reliable source for secondary aims. The Astra Gemini laser system has been operating at a high repetition rate for high intensity (0.5PW) experiments for a number of years and the Central Laser Facility has developed a target methodology to deliver to the user community the maximum number of solid targets and to fully utilise the available time on the laser. Targets for the High Accuracy Microtarget Supply (HAMS) system have been tested and have been proven to survive in a manner to allow shot rates comparable with the available laser repetition rate (0.1Hz). Investigations into target geometry have been carried out and debris production has been studied by high frame rate camera imaging. The study of the relationship between target geometry and debris production has allowed the design of optimal target support infrastructure, such as aperture size and structure, for high rep rate experiments on the Gemini system.

1. Introduction

The Target Fabrication Group at the Rutherford Appleton Laboratory has been delivering targets to the user programme in the UK for over 40 years. During this time there have been a wide range of targets delivered from simple thin foils through to complex multi-element, multi-material assemblies. However, one constant within this area has been the relatively low repetition rate of the laser which has limited the number of targets that have been needed for each experiment. It has been the case that for a standard experiment on the Vulcan laser system approximately 100-200 targets are needed. These can be varied from simple commercially available foils which are cut to size, through to complex 3D geometries such as AFI cones [1], shock experiments with backlighter targets [2] and gas cell targets as recently delivered on the Orion laser system at AWE [3]. The commissioning of the Gemini laser system [4] was a step change for target delivery at RAL with experiments requiring thousands of targets for each campaign. During the lifetime of the Astra Gemini system there has been an evolution to applications-based experiments on the facility. Some experiments are gas-jet based and are focussed on imaging of samples using x-ray and betatron beams and other experiments are focused on ion acceleration [5] and use solid targets in the range of 5 nm to a few µm. These require a mounting support and also a method of insertion into the laser focus and a refresh rate that is comparable to the laser. Initial experiments allowed for the targets to be made and mounted onto a standard target wheel with windmill target mounts. This type of target can be easily delivered and can provide thousands of targets per experiment [6], but are best suited for foils that are relatively simple to produce and that are thick enough to survive the shock and debris from a laser interaction. The drawback of such a system is that it is inherently low accuracy and for a laser system such as Astra Gemini that has a Rayleigh range of a few µm the target has to be realigned every shot using a focal spot camera that has to be driven into position. This slows the shot rate dramatically and reduces the utilisation of the laser facility whilst also putting a large assembly burden on the Target Fabrication resource that is available. This is clearly not useful for both science and applications-based experimental runs and also applications-based facilities may require large volumes of high specification targets that conventional techniques are not equipped to deal with.

The latest fabrication techniques have been utilised and are being integrated with the HAMS system that is being developed at RAL[7]. The debris that is generated from a high power (0.5PW) interaction with a thin foil target has been studied as well as the shock damage and debris mitigation techniques to deliver a robust and integrated solution to the delivery of high power laser experiments on ultra-thin solid targets for an F/2 focusing laser system.

2. The HAMS Concept for Targetry

The system aiming to fully utilise the high repetition rate of modern laser facilities which is being developed at the Central Laser Facility has been designed as a fully integrated target solution. It encompasses all elements from target manufacture through to the mounting of the target onto the delivery system. Accuracy of the delivery system is a key aspect as well as the integration of an on-line characterisation system for foil position, negating the need for a focal spot camera and reducing the turnaround time between shots. The baseline design for the system was to be able to run at one shot per minute which is an order of magnitude increase above the current standard for the Gemini system (on complex solid targets). There are five integrated areas of the HAMS system and these are; 1) the mechatronic stages, 2) the target interface, 3) the targets and their manufacture, 4) the interferometric position characterisation system and 5) the software integration of the whole system. The individual physical elements (minus software) are shown in Figure 1.

Figure 1. The 4 physical integrated parts of the HAMS system, which will allow for automated target alignment after integration of feedback programming

2.1. Mechatronic Stages

The HAMS stages consist of an X, Z linear translation stage with a tripod attached to the top to allow the stages to reach a working envelope of +/- 47.9mm in X and Y, 0-24mm in Z with an approximate +/- 10° tip tilt movement. A rotation mount on the top of the stages allows the target to be positioned normal to the beam or at an angle up to 45 degrees. Due to the F/2 parabola on the Astra Gemini laser the positioning of a target was agreed to be within an accuracy of +/- 4 μm in the direction of the laser (Z), this was to ensure it is within the laser focus to obtain the maximum intensity. A positional accuracy of +/- 10μm in X and Y was agreed to be enough for most targets. On top of all the translational stages a 360 degree rotary stage is required to rotate the wheel and deliver a new target to the focal position.

2.2 Interface Wheel
It is essential to be able to couple the targets to the mechatronic stages to a high accuracy to be able to fully utilise the high precision stage and targets. It is also essential that the interface is stable under experimental conditions. A wheel was machined in a ceramic that has good thermal stability and characterised to have flatness over the mounting positions of the target to better than 2 μm. This is within the tolerance that is acceptable for the wheel as long as other sources of error do not accumulate.

2.3 Target Sections
The HAMS system uses MEMS-based targets that are produced to the exact geometry for the interface wheel. It has been reported that targets can be made using MEMS techniques [8, 9, 10] and these techniques have been known for a number of years. This base technique is a key capability for the delivery of targets to high rep rate systems. It allows batch manufacture and allows standard processes and geometries to be used repeatedly at a relatively low cost per target. The targets can be designed to allow a number of different geometries that will be discussed. Target materials can be deposited onto the wafers before the fabrication process and different sections can be added to the wheel to allow thickness, geometry and material scans during a single pump cycle. Up to 8 target sections can be mounted on an interface wheel and flatness across the section and between the sections can be measured. Detailed characterisation of one section shows that it is flat to less than 0.5 μm and 2 sections mounted opposite each other on the interface wheel are flat to within 2 μm [11].

2.4 Target Alignment
The Gemini target area currently employs alignment systems that use retro or focal spot cameras. These limit the shot rate due to the fact that imaging the rear surface is very difficult on a thinner target as it becomes more transparent and a manual process for this can take a number of minutes. The CLF is developing a multi-wavelength interferometer system [7] which uses a focussed beam onto the target to allow the positioning of targets with a higher accuracy in Z. The system was not tested in this initial debris and survival test experiment and is not within the scope of this paper but will be fielded on further tests in the future. Work is being undertaken to implement a feedback loop which will eventually be able to automate the alignment of the targets.

3. Experimental Testing

3.1 Debris and Shock Damage Experiment Setup
The first experiment carried out in January 2016 was mainly focused on the understanding of the damage mechanisms for targets with the aim of benchmarking a design for future experimental campaigns and enabling the maximum utilisation of targets. The targets were shot using the F2 parabola at energies in the range of 11-15J which is typical for a Gemini experiment. The spot size was smaller than 10 μm which gives a minimum intensity on target of $3x10^{20}$ W/cm^2. Plasma mirrors were used [12] to give good contrast and the parabola had a pellicle to protect from damage. The main diagnostic for the damage experiment was optical imaging of the targets before and after the laser shot and the debris was investigated using a Photron FASTCAM SA-Z [13] which was used to image the interaction. The interaction was backlit with a continuous laser with wavelength 532nm and imaged for time-resolved shadowgraphy of the target dynamics. Typical frame rates for debris measurements were 100k to 210k frames per second (fps) dependent on the field of view to be imaged. A simplified chamber layout is shown in figure 2.

IOP Conf. Series: Journal of Physics: Conf. Series **1079** (2018) 012014 doi:10.1088/1742-6596/1079/1/012014

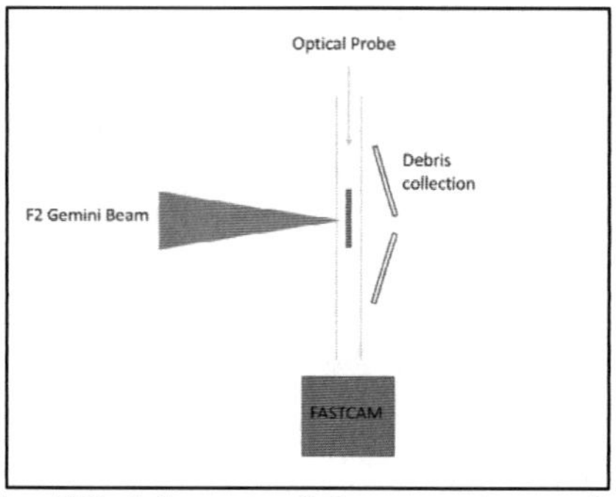

Figure 2. A simplified layout of the experiment for debris shots showing the F2 beam and the optical probe across the target.

A range of target geometries were tested to determine if shape, size and spacing of the targets was crucial. These are described below.

Table 1. The range of target geometries that were tested for the damage and debris studies.

Target Number	Aperture Geometry	Aperture Size
1	Circular	750um, 500um, 1000um, 2000um
2	Square	750um, 500um, 1000um, 2000um
3	HAMS sector	300um
4	Silicon 'Wheel'	2000um

In this experiment the first targets tested were array targets—not HAMS sectors—which contained circular and square targets with hole sizes of 500 µm, 750 µm, 1000 µm and 2000 µm with the spacing of the targets center to center being constant across a 25mm x 25mm, 325 µm thick silicon chip. The target material for these chips was 100 nm Silicon Nitride (SiN). The square targets were produced using wet etching techniques and the round targets were produced using Deep Reactive Ion Etching techniques. These targets were modified to have various protection layers on the front and the rear as the experiment progressed, either using thick metal supports to isolate the individual targets or thin glue layers in between the targets. In addition to this the final design for the HAMS sectors— the target that fits on the interface wheel—was tested. The HAMS targets have smaller apertures (300 µm) and are closer together than the aforementioned targets but are the same material (100nm SiN) and the sector is the same thickness (325 µm). The final target tested was a full 100mm silicon wafer with windows around the outside to replicate a full wheel. This target was 1 µm Parylene, poly(p-xylyene), suspended across a larger window. These targets are listed in Table 1 and are shown in Figures 3 to 5 where all chip sizes are 25 x 25mm.

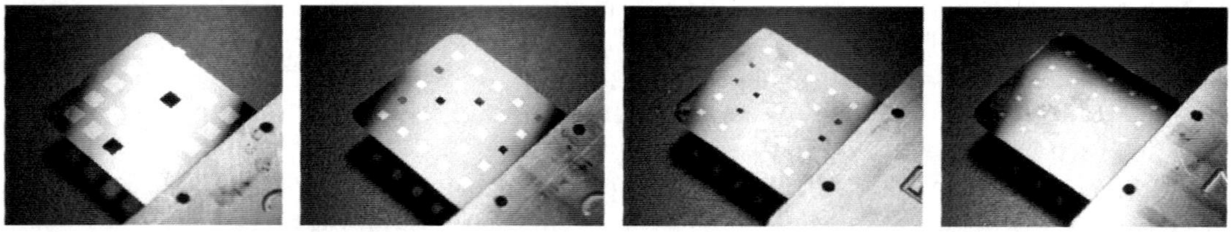

Figure 3. Pre-shot target arrays with different aperture sizes – some targets popped in manufacture. These were mounted to a standard target wheel with 8 array positions. Both circular and square versions of target foils in this geometry were manufactured.

IOP Conf. Series: Journal of Physics: Conf. Series **1079** (2018) 012014 doi:10.1088/1742-6596/1079/1/012014

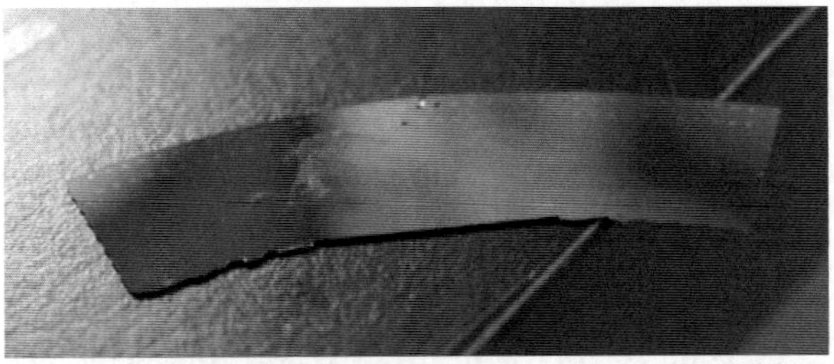

Figure 4. A HAMS sector produced by wet etching, which gives a rougher edge quality for the sector as it etches crystal planes. The 300 μm apertures can be seen in two rows patterned around the top edge of the sector.

Figure 5. A full silicon wafer patterned with 2000 μm aperture holes around the circumference.

3.2 Damage Results – Protection Study

Targets were photographed before and after shots with the laser to catalogue damage from the shot, due to the large amount of images these are all shown in appendix A. In most cases multiple shots were taken on the target arrays if there were surviving targets to determine the maximum number of shots possible before the array was spent.

The first targets tested were square array targets with 750 μm square holes in a 5x5 array that were protected only on the laser side with a large aluminium support. The support was 1mm thick with 1mm diameter tapered holes aligned to the targets to allow the laser beam in the aperture without clipping the mount. Before the shots there was a full complement of targets available in the array and 4 shots were taken in the corner positions. After the 4 shots there were only 5 surviving target positions (shown Appendix A-1), corresponding to a damaged target percentage of 76%

The second run of tests, again onto a 750 μm square target, were taken on arrays with an aluminium protection on the rear as well as the laser side. These shots were taken over two arrays to look at the target damage after a single shot as well as multiple shots as it was possible for the multiple shot run to shoot adjacent targets. In the single shot test, with a possible 22 targets available prior to the shot there were 18 still available after the interaction (shown Appendix A-2). In the multiple shot case, after 5 shots there were 11 targets remaining, from a total of 22 targets available before (Appendix A-3). This shows a damaged target percentage of 14% and 35% respectively.

The third run of shots were also taken onto a 750 μm square target and involved an array that had an aluminium protection on the front, similarly to the other targets but had glue lines between the targets on the rear side, acting as a shock propagation barrier as shown in figure 6. A total of 3 shots onto this array were taken after which there were 15 targets left (shown in Appendix A-4) leading to a damaged target percentage of 25%

Figure 6. A target with a glue protection layer on the rear.

3.3 Damage Results – Target Size/Geometry Study

Samples were then investigated for the size dependency for the survivability of the membranes. As described previously, targets of 750, 500, 1000, and 2000 μm apertures were prepared with a membrane thickness of 100nm SiN. The spacing of the aperture centres was kept at a standard distance in all cases at 3.5mm. All targets (both square and round geometry) were prepared with a glue protection layer on the rear surface as used for the targets in appendix A-4.

A method was devised to calculate the damage percentage for each target type. This takes into account the total available targets (pre-shot) on the array, the amount damaged specifically by the laser shots and normalising to the average number of shots taken on each array. Images of the targets before the experiments allowed verification of the number of potential targets to be shot Tp. The number of shot targets is classified as Ts, the number of targets available for shooting after a certain number of shots on the array is defined as Ta and the damage level from a scale of 0 to 1 (no damage to all targets damaged) is classified as

$$D = \left(\frac{(Tp - Ts - Ta)}{Tp - Ts}\right)$$

Normalising this to 4 shots, which was the average number of shots per array was possible by taking the ratio of the number of shots taken on the array compared to the average of 4. Therefore the normalisation factored damage percentage is

$$Dn = \frac{Ts}{4} D$$

The normalised target damage is thus classified as

$$Dn = \frac{Ts(Tp - Ts - Ta)}{4(Tp - Ts)}$$

Therefore

$$Dn = \frac{Ts}{4}\left(1 - \frac{Ta}{Tp - Ts}\right)$$

The geometry for the first targets was square, prepared using wet etching and the centre to centre spacing of the targets was constant at 5mm. By using the damage ratio calculation above, the percentage damage can be defined. A 2000 μm targets array was shot 4 times and this showed a damage of 68%, two arrays of 1000 μm targets were shot with damage levels of 89% and 87.5%, a 750 μm array was shot with a damage level of 33% and a 500 μm array was shot with the same 33% damage level. These targets are shown in appendix B-1.

These shots were then replicated on targets that were produced with round apertures using dry etching techniques. These targets were again 100nm Silicon Nitride but there was also an oxide layer in the target that stopped the etching process from damaging the target. This was approximately 1 μm in thickness and would not have been significant enough to prevent the laser from interacting with the target. This is verified with 1 micron shots described later in this paper. A 2000 μm target array was shot 5 times and this showed a damage level of 15%, a 1000 μm target was shot 3 times and showed a level of 9%, a 750 μm target was shot 3 times and showed 18% damage and a 500 μm target was shot 3 times and showed 0% damage of neighbouring targets. These targets are shown in appendix B-2.

The graph in Figure 7 shows the results for both the square and round targets given in percentage terms and the results normalised for 4 shots. Error bars of +/- 10% are used, this is due to the amount of targets that are able to be damaged as nearest neighbours depending on where the target is shot. If it is shot in the corner then there are 3 surrounding targets, or 12.5% of the remaining array (25 minus the shot target) if the target is shot directly in the centre then there are 8 surrounding targets with a potential to be damaged, or 33% of the remaining targets. There is therefore is a maximum 20.5% variation in the damage potential for each shot and this reduces after each shot on the array as targets are destroyed.

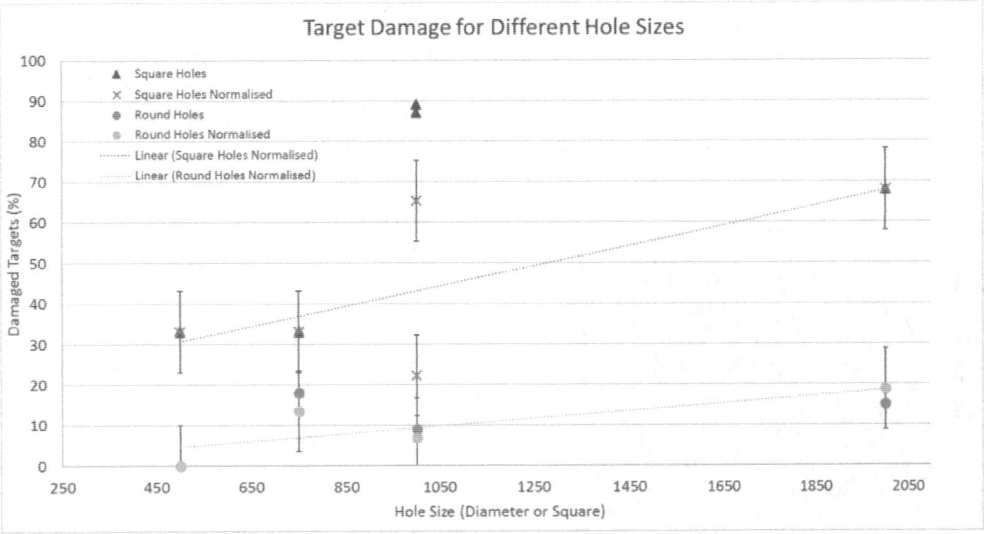

Figure 7. The target damage levels for both round and square holes and for different hole dimensions with constant spacing center to center of 3.5mm.

3.4 Rotational Targets

Two types of targets were tested that would be suitable for fielding on the final HAMS system, these both have rotational symmetry. The first design is based on the sector geometry as previously discussed with pre-shot imagery shown in Figure 4 and post-shot in Figure 8 below (center and left). The design was processed using wet etching and comprised 300µm square apertures with a pitch of 1000µm, the target material consisting of a 100nm thick Silicon Nitride membrane. The second target design was a full wafer with etched holes around the circumference shown in Figure 5 and Figure 8 (right) before and after shots, respectively. This target has 1 µm Parylene films suspended across the aperture.

Figure 8. Radial targets in both HAMS sector (left) and full wafer geometries (right) with the beam footprint highlighted (center).

The shots on the wafer segment (left) showed a laser footprint that was of the order of 2mm diameter (center). This relates to a FWHM focal spot size on the Gemini laser of 2.5 μm, and this can be optimised using adaptive optics [14]. An image of the focal spot is shown in figure 9 with a lineout of the focal intensity. The focal intensity drops off to a level that there should be no appreciable laser energy at the 2mm diameter above the ablation threshold, however damage is seen. Currently the most likely source for this footprint is amplified spontaneous emission (ASE) leaking through the laser pinholes but more investigation is needed on this. Four shots were taken on this target segment and it is clear on these shots that the adjacent targets were damaged by the footprint. It is believed that there was enough energy in the ASE footprint to pop the targets that were adjacent to the interaction but this needs further investigation. However, it is also clear that for a small aperture on these targets it was only possible to shoot every third target. For the larger targets with a 1 μm Parylene window the beam footprint is kept within the area of the target and is not seen. A number of shots were taken on this target on adjacent windows and whilst there was a good survival of nearest-neighbour targets the debris that was generated from one of the shots was substantially more than for other shots and popped the pellicle which was protecting the parabola.

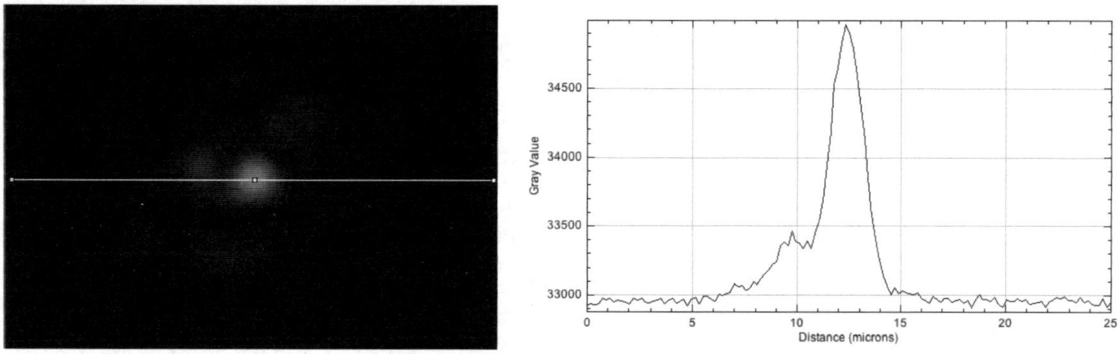

Figure 9. An image of the F2 focal spot and a lineout of the intensity distribution

Images of the debris created from these shots are shown in figure 10, again the silicon wafer seen is 325 μm thick and this is at 0.1ms after the shot, It can be seen for shot 39 there is much more material that is present at the same time than for the other targets. We believe this is the most likely cause for the damage to the pellicle as the alignment was not ideal on this shot and we may have irradiated the mount with more laser intensity than other shots.

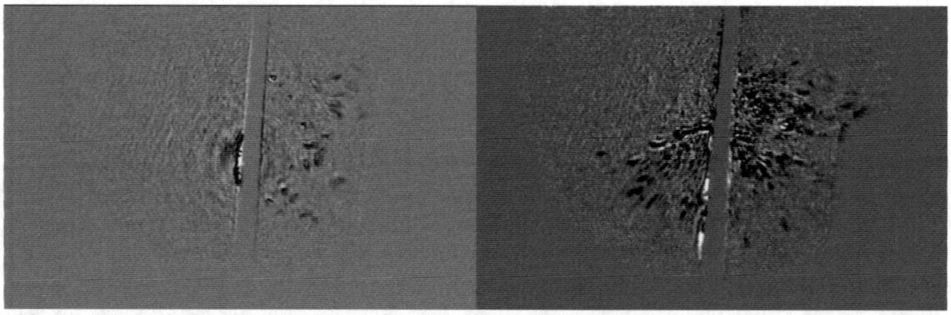
Figure 10. Debris measurements for the same time after shot for 2 identical targets.

3.5 Debris Measurements

As described in Section 3.1, debris measurements were taken using a FASTCAM system to allow for the tracking of the particles that were ejected from the target. The debris that was generated varied due to the target geometry as well as the target protection and separation layers that were present. In all cases

IOP Conf. Series: Journal of Physics: Conf. Series **1079** (2018) 012014 doi:10.1088/1742-6596/1079/1/012014

the raw images have been processed to remove the background and to enhance the appearance of the debris particles for visualization. The aperture of the target is centered in the image in all cases and although hidden by the silicon chip bulk material is seen from where the plasma plume emanates.

Debris measurements were taken on square targets with 2 mm apertures, glue protection on the rear and no front protection, shown in Appendix B-1 (2 mm square apertures). The data shown in Figure 11 was taken at 100k FPS and captures the initial plasma blow-off from the laser side in frame 1 (0.01 ms) which is quickly followed by particulate at 0.1-0.3 ms with the remaining target material being driven back towards the laser as the target relaxes at approximately 1 ms from the interaction.

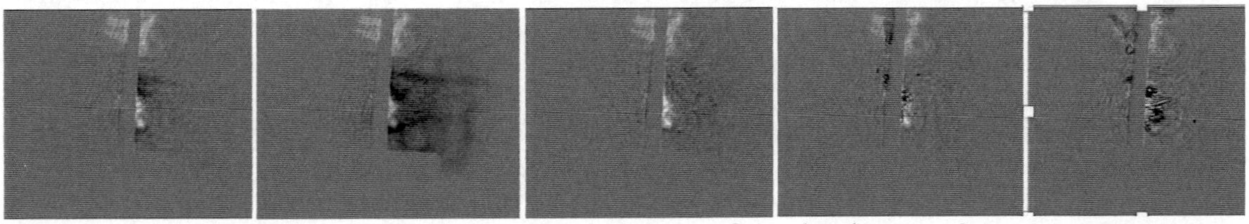

Figure 11. FASTCAM data at t=0, 0.01, 0.1, 0.3 and 1 ms from the shot on a 2 mm square aperture.

Comparison of this data with a smaller aperture target shows a significant reduction in the large scale debris that is seen in footage taken 0.1 ms post shot. This is most likely due to there being a smaller amount of target material that is left un-ablated to relax back after the shot and break away from the mount. The images below in Figure 12 show a 1 mm square target at the same time periods after the shot. It is noticeable that in images for 0.3 ms and 1 ms there is no apparent target material and at t = 0.1ms there is little large scale debris present.

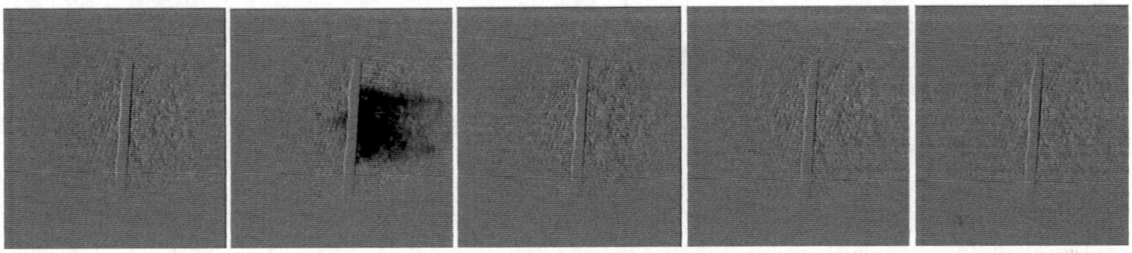

Figure 12. FASTCAM data at t=0, 0.01, 0.1, 0.3 and 1 ms from the shot on a 1 mm square aperture

Comparing the data from a shot on a 2 mm round aperture target as described and shown in Appendix B-2 (2 mm) with targets in Figures 11 and 12 shows again that there is less debris after the initial shot than for a square target. Through use of particle tracking software, the trajectories of specific debris can be determined and are highlighted in images 4 and 5 of Figure 12. This particle travels 2098 μm in 81 ms (which can be calculated using the known width of the silicon wafer in the image which was measured externally to the experiment as a calibration). This equates to a minimum particle ejection speed of ~2.5 m/s for large particles depending on their trajectory relative to the focal plane, resulting in a higher speed. It is also possible to calculate the speed of the particles that are initially ejected by referring to Figure 13, image 2; the particle cloud disappears from the field of view within 2 frames. Taking the width of the image as 5 mm these particles have to travel 2.5 mm in 20 μs which equates to an average speed of ~125 m/s.

IOP Conf. Series: Journal of Physics: Conf. Series **1079** (2018) 012014 doi:10.1088/1742-6596/1079/1/012014

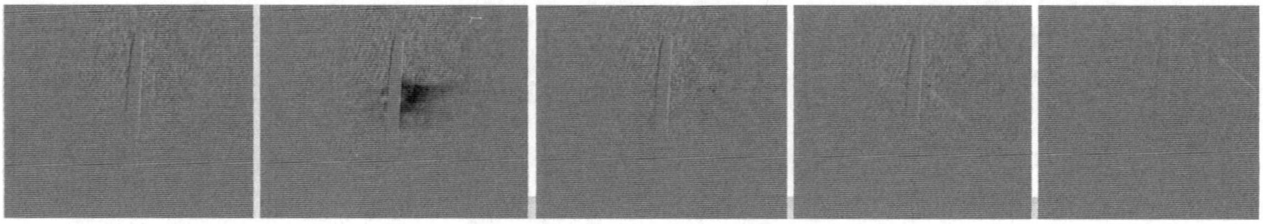

Figure 13. FASTCAM data at t=0, 0.01, 0.1, 0.3 and 1 ms from the shot on a 2mm round aperture

In all of the above cases it is clear that there is debris generated from different mechanisms when a target is ablated; initial plasma blow off occurs at very short timescales; small particle production, mainly travelling back towards the laser which takes approximately 0.1-0.3ms to travel a few mm; and the larger scale debris, which is primarily the shrapnel from the support or the foil itself that propagates over a longer timescale. This larger debris is of concern as the shrapnel may damage the parabola and optics, whereas the smaller and faster debris could cause issues with the coating of optics or contamination of other targets. Experimentally it is desirable to reduce debris wherever possible.

4. Conclusions

Detailed studies have been carried out on the size and the geometry of the targets and a clear idea of the ideal target parameters for the Gemini system has been ascertained. It can be seen there is a clear trend that a smaller target aperture increases the chance of neighbouring targets surviving on the array. There however would be a lower limit to this which would be characterised by the geometry of the incoming laser beam and the thickness of the target mount. In these examples for a F/2 beam through a 325μm thick wafer the hole size would need to be 325μm square to stop the beam clipping the target at normal incidence. It also seems preferential to use circular rather than square aperture targets. This may be due to the increased stresses or shock-focussing into the corners of the square target apertures. Further work needs to be carried out to fully understand this hypothesis. Debris studies carried out show that smaller target apertures create less large-scale debris and therefore are preferable for high rep rate experiments where the build-up of such debris will degrade the system.

It is also noticeable that where there is protection in the front and rear of the targets, in the case of this investigation using plates or glue lines there is a reduction in the damage to the surrounding targets. This could be down to the shadowing of the aperture from some of the plasma and debris, or it could be some impedance change that stops shocks propagating across the surface of the holder. Further work is needed to investigate this process.

Further work is planned to test the targets on the complete HAMS system and to verify the high repetition rate of the rest of the integrated technology, however the experiment carried out shows that MEMS targets are applicable to high rep rate operation of systems such as the Gemini laser with the appropriate target protection and therefore the cost effectiveness and the reproducibility of such targets can be exploited. The integration of the interferometer to aid alignment is being developed to verify target position and to introduce a feedback loop for alignment.

IOP Conf. Series: Journal of Physics: Conf. Series **1079** (2018) 012014 doi:10.1088/1742-6596/1079/1/012014

References

[1] Kodama R et al 2001 *Nature* **412,** 798-802
[2] Hawreliak J et al 2012 *AIP Conference Proceedings* **1426**, 975; doi: 10.1063/1.3686440
[3] Spindloe C et al 2017 *High Power Laser Science and Engineering* **5** e22
[4] Hooker C et al 2006 J. *Phys. IV France* **133** 673-677
[5] Snavely R et al 2000 *Phys. Rev. Lett*, **85**, 2945
[6] Spindloe C, Dival E, Schaumann G, Nazarov W and Tolley M K 2009 *Central Laser Facility Annual Report* 301-305
[7] Booth N et al 2014 *Proc. SPIE 9211, Target Diagnostics Physics and Engineering for Inertial Confinement Fusion III*, 921107
[8] Spindloe C *et al* 2016 *J. Phys.: Conf. Ser.* **713** 012002
[9] Zaffino R et al 2018 *Microelectronic Engineering*, **194**, 67–70
[10] Anastasi P.A.F., Burge R.E. 1992 *X-Ray Microscopy III. Springer Series in Optical Sciences*, **67**. Springer, Berlin, Heidelberg
[11] Astbury S et al (Submitted) *CLF Annual Report 15-16*
[12] *Streeter M et at 2008-2009 CLF Annual Report* 225
[13] https://photron.com/fastcam-sa-z/
[14] Symes D et al 2015 *Central Laser Facility Annual Report Article 55.*

IOP Conf. Series: Journal of Physics: Conf. Series **1079** (2018) 012014 doi:10.1088/1742-6596/1079/1/012014

APPENDIX A

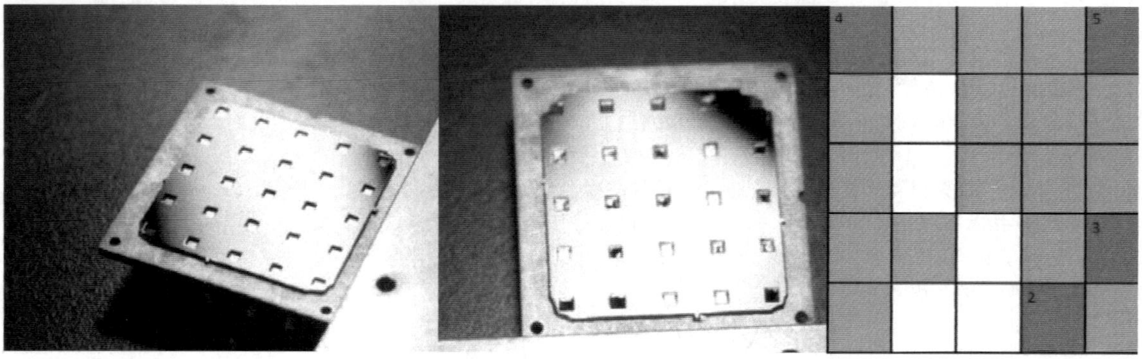

A-1. (Left) The target before positioning in the chamber showing a full complement of foils and after the shots (center) showing a large amount of damaged foils and (right) the shots taken on the target with red indicating shot, grey indicating damaged and white indicating surviving.

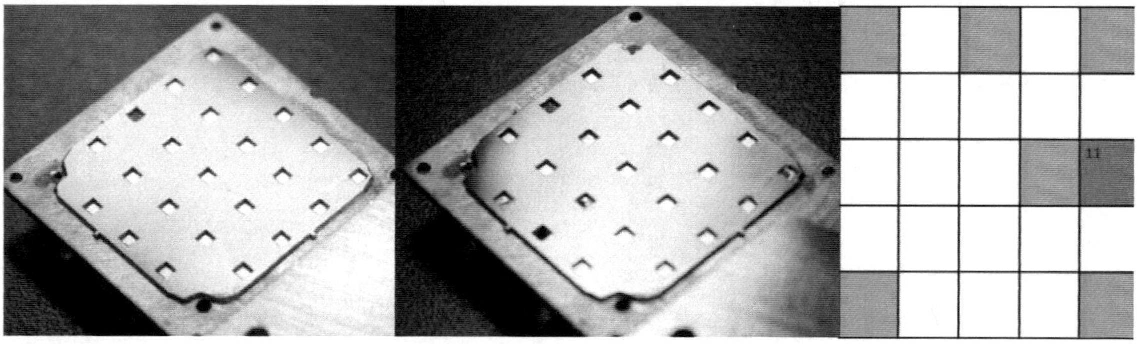

A-2. (Left) The target before positioning in the chamber (with the Al protection removed) showing a 22 out of 25 foils available (center) showing a shot and damaged foils and (right) the shots taken on the target with red indicating shot, grey indicating damaged and white indicating surviving.

A-3. (Left) The target before positioning in the chamber (with the Al protection removed) showing a 22 out of 25 foils available (center) showing a the shot and damaged foils and (right) the shots taken on the target with red indicating shot, grey indicating damaged and white indicating surviving.

A-4. (Left) The target before positioning in the chamber showing 23 out of 25 foils available and after the shots (center) showing the damaged foils and (right) the shots taken on the target with red indicating shot, grey indicating damaged and white indicating surviving.

APPENDIX B

B-1. (Left) Square target shots from 2000 microns, 1000 microns x 2, 750 microns and 500 micron apertures. Target damage can be easily seen on the foils.

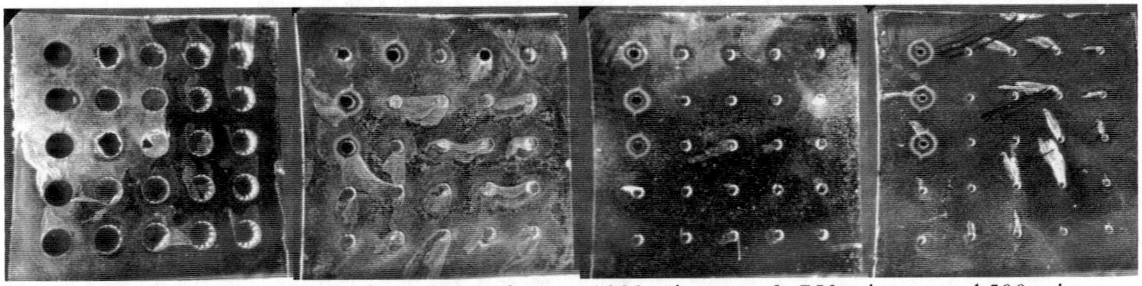

B-2. (Left) Round target shots from 2000 microns, 1000 microns x 2, 750 microns and 500 micron apertures. Target damage can be seen on the foils.

IOP Conf. Series: Journal of Physics: Conf. Series **1079** (2018) 012014 doi:10.1088/1742-6596/1079/1/012014

APPENDIX C

C. Image of the F2 focal spot with a FWHM of 2.5 microns

IOP Conf. Series: Journal of Physics: Conf. Series **1079** (2018) 012015 doi:10.1088/1742-6596/1079/1/012015

Characterisation of the oxide effects on aluminium opacity targets

D Haddock[1], S White[2], C Spindloe[1], D Riley[2] and M Tolley[1]
[1] Science & Technologies Facilities Council, Rutherford Appleton Laboratory, Harwell Campus, Chilton, Didcot, Oxon, OX11 0QX, UK

[2] School of Mathematics and Physics, Queens's University Belfast, BT7 1NN, UK

Email: david.haddock@stfc.ac.uk

Abstract. 900 nanometre thick aluminium strips were produced by the Target Fabrication Group for an experiment at the Central Laser Facility. They were specified to be supported horizontally at either end across an 800 micron void and to be 500 microns in height. The production method used a Parylene (C_8H_8) supporting layer which was removed by oxygen plasma etching. This process resulted in a problematic contaminant layer that caused substantial increase to the absorption of X-rays in the laser campaign. The contaminant layer was characterised by optical microscopy and Scanning Electron Microscopy. Energy Dispersive X-ray Spectroscopy was able to detect a high relative abundance of carbon and oxygen in the plastic etched strips when compared to control samples. It has been shown that partially etched Parylene can form carboxylic groups with a ratio of 4:1 carbon to oxygen atoms. This effect was observed and exacerbated by the sample geometry not in sufficient contact with the grounded chamber baseplate causing charging and therefore insufficient hydrogen abstraction in the plasma. An alternative production method produced strips that were not as flat over the open area as the previous method but had no significant contaminant layer. This process involved coating a release layer of sodium chloride before the aluminium strips. When lowered into water, the salt dissolves in water and the strips are suspended on the surface tension, allowing the strips to be picked onto the mount. Each target required measuring for form using a white light interferometer where around half were within specification.

1. Introduction

This paper presents analysis of an aluminium strip target made by the Target Fabrication Group at the Central Laser Facility (CLF). The production method is outlined alongside characterisation data of the foil flatness and relative abundance of contaminants present on the surface. An alternative production method is presented to reduce contaminants and its suitability is discussed.

Techniques used to make the strip include coating plastic by Chemical Vapour Deposition, Aluminium coating by Physical Vapour Deposition and oxygen plasma etching; analysis techniques include Light Microscopy with Dark-Field Illumination, Scanning Electron Microscopy with Energy Dispersive X-ray Spectroscopy and White Light Interferometry. All resources used are part of the Target Fabrication capabilities at the CLF.

1.1. Target Specification

An experimental campaign undertaken in January 2017 on the Vulcan laser system required targets of aluminium 900nm thick, 500μm high free-standing over a frame with 800μm gap. The form of the strip was specified to be as flat as possible and have no angles greater than 5 degrees with respect to the mount plane. A sketch of the design is shown in Figure 1.

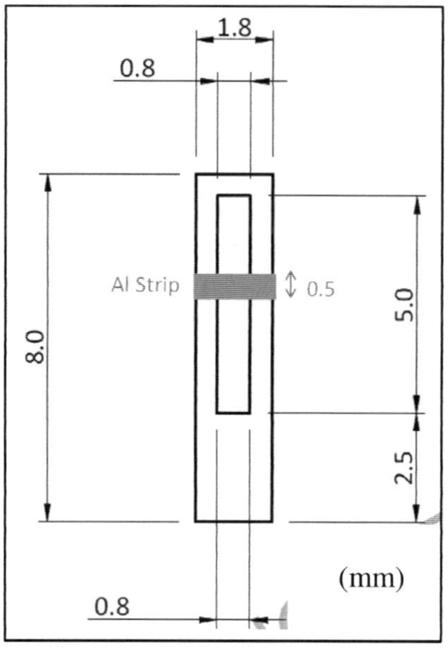

Figure 1. Strip target specification sketch. All dimensions in mm. The aluminium strip is 900 nanometres thick. The supporting structure was made from photo-etched steel

2. Initial manufacture method

In order to easily place and support an Aluminium strip whilst also achieving flatness across the area the chosen manufacture method included supporting the strip with a plastic under layer that was removed by oxygen plasma etching.

2.1. Strip manufacture

A coating of Parylene N (trade name for a variety of chemical vapour deposited poly(p-xylene) polymers) of thickness 5 microns was deposited onto a standard microscope slide by Chemical Vapour Deposition. The microscope slide is then coated with the desired thickness of aluminium by thermal evaporation in a thin-film deposition plant. The "strip" width of 500μm was defined by using a mask made of 100μm thick photo-etched copper. Placing the mask flat down onto the Parylene coated microscope slide allowed for a coating with little shadowing as tested by a surface profilometer. The deviation in thickness coated across a microscope slide is considered well within reasonable error margins (±5% deviation from total thickness).

IOP Conf. Series: Journal of Physics: Conf. Series **1079** (2018) 012015 doi:10.1088/1742-6596/1079/1/012015

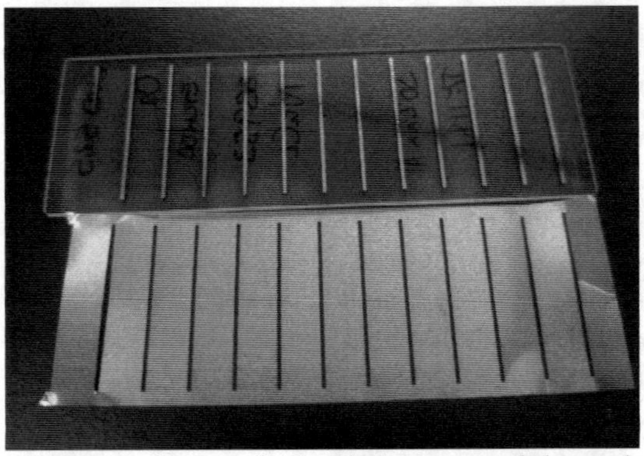

Figure 2. Strips of Aluminium coated through the copper mask onto a microscope slide. The strip edges were well-defined with little shadowing

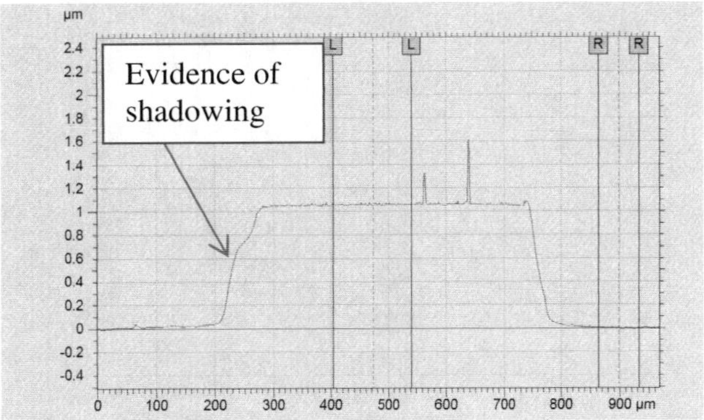

Figure 3. Touch probe scan of the step height across a strip as shown in Figure 2. The difference of the two "L" and "R" zones measures 1.079(±0.064µm) and this is taken to be the thickness of the coating. The amount of shadowing in coating can be measured by looking at the quality of the "top-hat" profile

The touch probe scanned across the width of each strip and resulted in the scan in Figure 3. This profile measurement also serves as the thickness measurement for the coating run.

2.2. Oxygen plasma removal of plastic

Individual strips are cut from the slide manually using a scalpel and glued over the open area of the mount using a small amount of epoxy glue. This is placed inside a Plasma-Enhanced Chemical Vapour Deposition plant where oxygen plasma removes plastic by Hydrogen abstraction[1]. A flow rate of 30sccm oxygen is maintained at a pressure of 100mT by a gate valve to a turbo vacuum pump. A power of 150W is maintained by a discharge between two electrodes, creating plasma. This Oxygen plasma removes plastic at a rate of ~100nm/min. Choosing a time of >50minutes ablates ~ 5microns of plastic. This method has been verified by taking a control substrate (glass microscope slide) with plastic and etching under the same conditions. No remaining plastic is detectable by surface profileometry. The sample is then taken out and visually inspected to ensure all plastic is removed, the dimensions fit that of Figure 1 and there is no visible debris or deformity. The mount with the strip can then be placed onto a CLF standard aluminium post ready for laser irradiation in the target area.

IOP Conf. Series: Journal of Physics: Conf. Series **1079** (2018) 012015 doi:10.1088/1742-6596/1079/1/012015

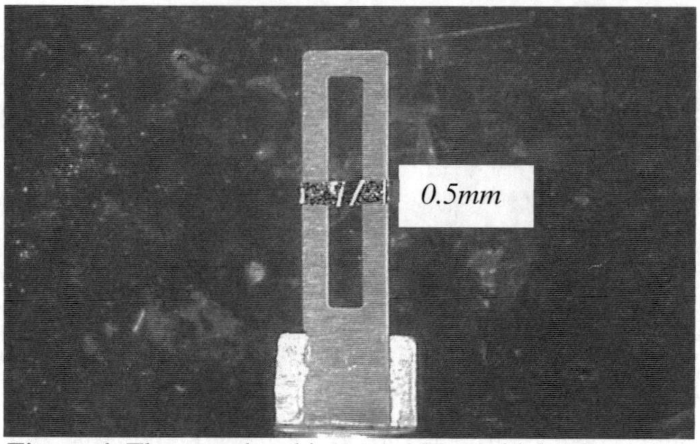

Figure 4. The completed item. A 900nm thick, aluminium Strip supported either end over a 0.8mm aperture

The result using this production method yields a flat and reproducible strip that can be batch produced. When measured by Interferometry all targets were within specification and could be produced with high levels of consistency.

3. Alternative production method

The production method was chosen to remove any source of plastic and include no Oxygen plasma etching during processing.

3.1. Float off procedure

Firstly a release layer of sodium chloride was deposited on a substrate by way of Physical Vapour Deposition in a thermal evaporation coating plant. Then, using the same strip mask as in the first production method, aluminium is coated through the mask to define the strips onto the sodium chloride. When the substrate is lowered into de-ionised water the salt dissolves and the aluminium is released from the surface and is suspended on the surface tension of the water. The strip can then be lifted out of the water directly on to the steel target mount. This process was done manually by an experienced fabricator, it was labour intensive as each strip had to be processed by hand and produced varying results in terms of the flatness of the strip across the length.

3.2. Form measurement using White Light Interferometry

Each strip had its form over the open area measured to test against specification; White Light Interferometry gave a surface scan that allowed the form to be assessed. Due to the strict specified criteria for form around half of those produced were rejected. Figure 5 and 6 show an interferometry scan of the surface profile with a line-out across its length.

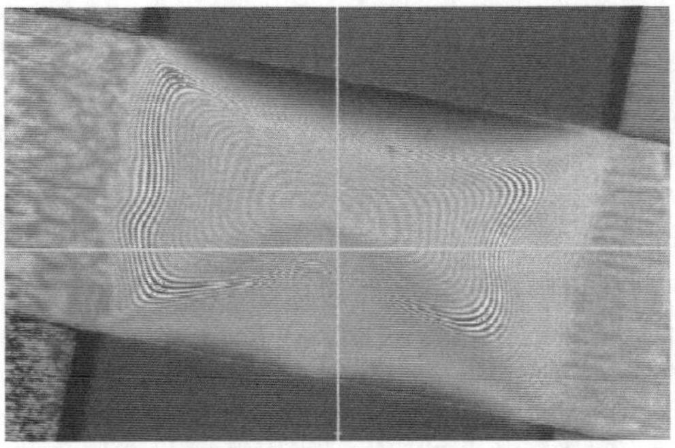

Figure 5. Interferometry of a target strip produced by float off. The interference fringes show the form of the strip

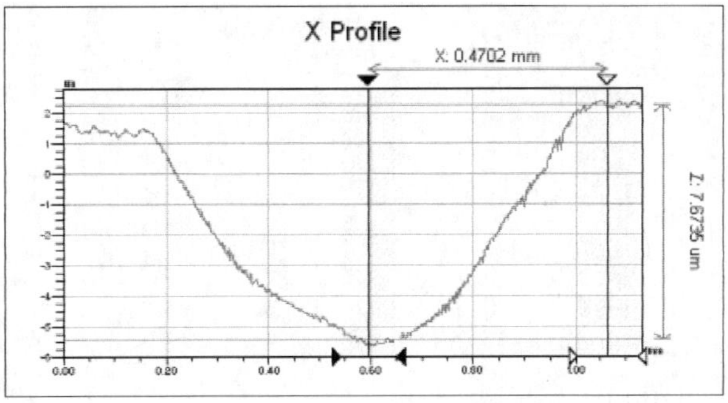

Figure 6. Resultant profile along the length of the strip, measured using a White-Light Interferometer. Note the scale is exaggerated. The strip falls 7 microns in the middle relative to the supported ends

4. Oxide observation by X-ray absorption

During the experimental campaign the users group noted an unexpectedly high X-ray absorption through the oxygen-etched sample strips, indicating there may be a much greater oxide layer than there would be natively in aluminium, Oxygen plasma can create a deep oxidation in Aluminium[5]. As a result, the user group took two identical shots: one on a standard production target and one produced via the alternative method. The results showed much greater X-ray absorption in the standard manufacture target, indicative of a thicker-than-native oxide layer. The X-ray absorption on the alternative produced target was considerably less and was characteristic of a native aluminium oxide layer.

The presence of the Oxide layer made the standard targets unsuitable for the rest of beam time. Target Fabrication undertook some retrospective characterisation research and chose the alternative production method for the remainder of the experimental campaign.

5. Contaminant characterisation

The produced strips were inspected with dark field and bright field optical techniques as well as Scanning Electron Microscopy (SEM) with Energy Dispersive X-ray Spectroscopy (EDX) for contaminant quantities.

5.1. Optical Imaging

Figures 7 and 8 are optical images taken of the strips after etching. Figure 7 is a bright field image and figure 8 is a dark field image. Figures 7 and 8 clearly show plastic remaining around the strip despite prolonged etching.

IOP Conf. Series: Journal of Physics: Conf. Series **1079** (2018) 012015 doi:10.1088/1742-6596/1079/1/012015

Figure 7. Bright field image of Aluminium strip produced by production including a CH layer removed by Oxygen plasma etching. There is some plastic visible at the top and bottom

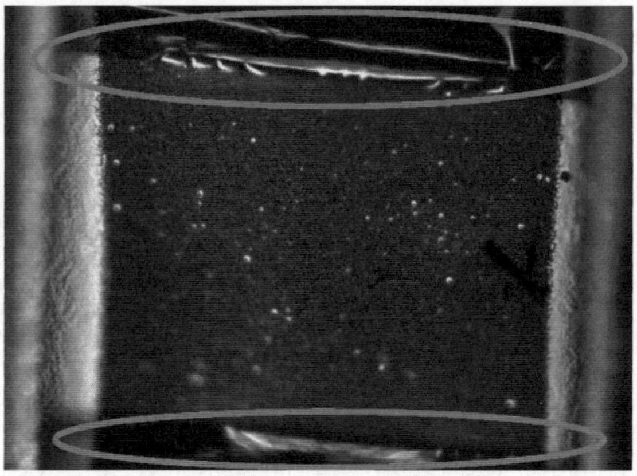

Figure 8. Dark-field image of the strip. Dark-field microscopy excludes un-scattered light from the final image. Under dark field illumination plastic containment can be seen clearly in areas at the top and bottom

5.2. Scanning Electron Microscopy (SEM) imaging

Imaging the extent of the plastic contaminant using optical microscopy is difficult due to its transparency but SEM allows imaging irrespective of optical properties.

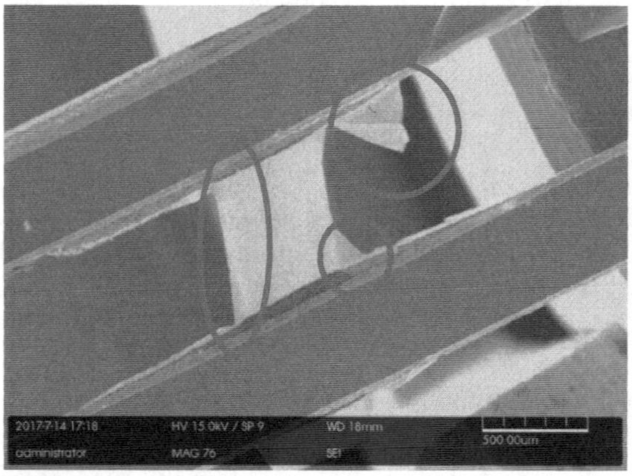

Figure 9. Scanning Electron microscopy image of the same strip as figure 5. Plastic can be clearly seen around the strip

Figure 10. Scanning Electron Microscope image of a strip produced by a method of "float off" this process includes no plastic coating or oxygen plasma processing. The strip is free of contaminants

Figure 9 shows the extent of the remaining plastic. It is clear despite prolonged exposure to the Oxygen plasma it is not being removed fully. For comparison the below image (Figure 8) is a strip made by the alternative production method.

5.3. Identifying the contaminant using Energy Dispersive X-ray Spectroscopy (EDX)

Energy-dispersive X-ray spectroscopy of strips with contaminant and without contaminant; can be used firstly to; rule out considerable high-Z contaminant, and secondly, to give a relative abundance of carbon and oxygen content and estimate the extent of the contaminant. (Only relative abundance of low-Z elements can be obtained using EDX)

IOP Conf. Series: Journal of Physics: Conf. Series **1079** (2018) 012015 doi:10.1088/1742-6596/1079/1/012015

Using a beam of 5KeV and choosing an area of ~50μm^2 in the centre of the strip a 100,000 counts spectrum was taken. The results for the samples in Figure 9 and Figure 10 are in Tables 1 and 2.

Table 1. EDX element data by relative abundance by weight and atomic number for a strip that had plastic removed by Oxygen plasma (Figure 9 and 10)

Element	Weight %	Atomic %
Al K	53.29	37.31
C K	27.52	32.50
O K	19.19	30.19

Table 2. EDX element data by relative abundance by weight and atomic number for a strip that had no plastic or plasma processing (Figure 10)

Element	Weight %	Atomic %
Al K	76.24	60.99
C K	8.27	11.16
O K	15.49	27.84

Both the strips contained no detectable high-Z contaminant. Relative carbon content is far higher in the strips with the plastic coating that has been subjected to oxygen plasma. This shows that there is plastic remaining on the surface of the strip as well as around the edges. It also shows a slightly higher abundance of oxygen, this implies that the oxygen is reacting chemically with the carbon to form carboxylic groups, this has been show to occur in partially oxygen etched Parylene[1]. Carboxylic groups are resistant to further reaction with oxygen plasma which suggests why some plastic remained on the surface.

5.4. Modelling of electron interaction area and X-ray volume

]. To estimate a thickness of contaminant on the surface of the aluminium a CASINO was performed as shown in figure 11. CASINO software is a "*Monte CArlo Simulation of electroN trajectory in sOlids*" specially designed for low energy electron interaction in a bulk and thin foil[2] CASINO calculates the electron trajectory by a series of separate scattering events, after which the electron is absorbed and energy is released as electromagnetic radiation. The depth of the electron "tear-drop" shape is dependent on a number of factors including density of material and starting electron energy. This technique has been used to estimate the thickness of thin film layers[3][4]. Using a simulated electron beam that is absorbed in a characteristic teardrop shape it is possible to analyse where electrons are absorbed and released as X-rays.

A 5KeV electron beam was modelled entering an Aluminium substrate with a Carbon/Oxygen contaminant on the surface. A 225nm contaminant layer of element ratios 4C:O:3H and density 1.06g/cm3 is modelled below. This ratio was chosen to represent Carboxylic acid. Different thicknesses of contaminant were trailed to match the EDX data, through trial and error; this could then be used to estimate the thickness. A 5KeV electron beam was modelled entering an Aluminium substrate with a Carbon/Oxygen contaminant on the surface. A 225nm contaminant layer of element ratios 4C:O:3H and density 1.06g/cm^3 is modelled below. This ratio was chosen to represent Carboxylic acid. Different thicknesses of contaminant were trailed to match the EDX data, through trial and error; this could then be used to estimate the thickness.

IOP Conf. Series: Journal of Physics: Conf. Series **1079** (2018) 012015 doi:10.1088/1742-6596/1079/1/012015

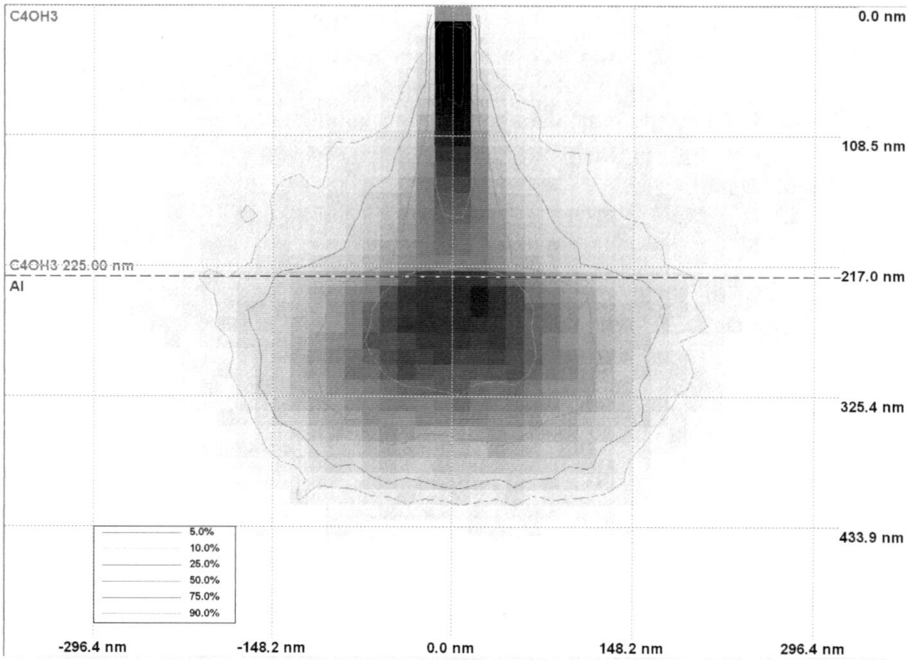

Figure 11. A simulation of 5KeV electrons in a 225nm plastic contaminant containing a ratio of 4C:O:3H and a density of 1.06g/cm^3 on top of solid Aluminium. The teardrop shape shows where electrons are absorbed and reemitted as X-rays.

Table 3. X-ray yield for each element present in the simulation as a percentage of total

Element	Weight %
Al K	57
C K	27
O K	16

Using a contaminant layer of 225nm in the simulation the abundance of element matches the data collected from the contaminated strip, suggesting the contaminant thickness is on this order of magnitude.

An EDX map of the strip would be advantageous however this was difficult to achieve because of the charging of the areas of plastic contaminant around the strip.

6. Conclusions

Self-supported Aluminium strips of 900nm in thickness and 0.5mm in width were batch produced by using a Parylene support film which were consistently flat across the open area as measured by interferometry. The oxygen plasma etch to remove the plastic support film introduced a contaminant layer of carbon and oxygen atoms which rendered them inadequate for experimental use. The contaminant layer was confirmed by EDX to be an abunance of carbon with some inclusion of oxygen.

Although chemical bonding could not be probed with the equipment it is theorised that the oxygen plasma chemically bonded with the plastic support layer and produced carbon-oxygen functional groups which were resistant to further etching. A 225nm 4:1 carbon-oxygen contaminant on the surface of aluminium was modelled under a 5KeV beam, the energy emitted as X-rays from each element was

IOP Conf. Series: Journal of Physics: Conf. Series **1079** (2018) 012015 doi:10.1088/1742-6596/1079/1/012015

comparible to data gathered from EDX of a contaminated strip. Increased contaminant occurred with these samples more readily than comparative samples because of the geometery of the sample in the plasma, poor conduction with the baseplate inside the process chamber caused the sample to charge.

As a result of problematic contaimant layers an alternative production was used. Coating the strips onto a release layer of sodium choride succeeded in producing a product with no significant contaminant, however, this was labour intensive and produced inconsistent flatness across the open area. The criteria for form required each one to be quality checked and only around half of those produced were suitable.

7. Further Work

To further analyze the surface chemical nature, X-ray photoelectron spectroscopy could be employed. This technique can be used to analyze the surface chemistry of a sample and would give insight into the oxygen containing functional groups present on the surface as a result of oxygen plasma etching.

References

[1] Callahan R, Raupp B, and Beaudoin S 2001 Effects of gas pressure and substrate temperature on the etching of parylene-N using a remote microwave oxygen plasma. *J. Vac. Sci. and Technol. B: Microelectronics and Nanometer Structures,* vol. 19, p725-731

[2] Gołda C M, Brzychczy-Włoch M, Faryna M, Engvall K, Kotarba A 2013 Oxygen plasma functionalization of parylene C coating for implants surface: Nanotopography and active sites for drug anchoring. *Materials Sci. and Eng. C* vol. 33, p4221-7

[3] Canli S 2010 Thickness determination of thin films by energy dispersive x-ray spectroscopy *a thesis submitted to the graduate school of natural and applied sciences of middle east technical university*

[4] Myint K, Barfels T, Kuhr J 1998 EDX depth profiling by means of effective layers *Fresenius J Anal Chem* vol. 361(7), p637-639

[5] Baier-Saip J, Babrera A, Zarate R 2005 Deep oxidation of aluminum by a DC oxygen plasma *Surface and Coating Tech.* vol.195 p168-175

IOP Conf. Series: Journal of Physics: Conf. Series **1079** (2018) 012016 doi:10.1088/1742-6596/1079/1/012016

Manufacture and Characterisation of Materials for High Power Laser Targets

A. Hughes[1], D. Haddock[1], W. Nazarov[2], C. Whyte, M. Tolley[1] and C. Spindloe[1]

[1]Science and Technology Facilities Council, Rutherford Appleton Laboratory, Harwell Campus, Chilton, Didcot, Oxon, OX11 0QX, UK

Email: aasia.hughes@stfc.ac.uk

Abstract. The Target Fabrication group at the Central Laser Facility (CLF) works to develop and produce novel targets for the high-power laser community. Targets range from simple foils to more complex geometries and with many components and key part of the research is materials development. We describe the recent advances in the CLF in surface preparation for electroplating and the manufacture of thin films. Characterisation of these materials is carried out using Atomic Force Microscope and Scanning Electron Microscopy. These advances have been applied to target components that have been used experimentally on the Vulcan Laser system at the Rutherford Appleton Laboratory.

1. Introduction

The Target Fabrication group in the CLF has been delivering targets to the UK high power laser community for over 40 years with thin films being made from the opening of the CLF in 1978. The Central Laser Facility established a dedicated chemistry laboratory to support its internal program in 2010 and has since then developed the capabilities to manufacture low density polymeric foams using critical point drying techniques, ultra-thin polymer films using dip and spin coating techniques [1, 2] and also electroplating capabilities for micro objects [3]. The focus of the research has been to push the limits of the capabilities to make materials for high power laser experiments through a use of novel techniques. Advances in the characterisation of these materials has given an insight into the microscale features of these target materials and has enabled a greater understanding of the limitations of the processes. This information is fed back to the user community to enable them to design technically possible targets that achieve their scientific aims and this collaboration enables the target fabrication group to focus its future research on the developing needs.

2. Electroplating

We have been developing electroplating capabilities to reduce the need for glue layers for multi-layered and multicomponent targets. Glue layers in such targets can have detrimental effects on the results obtained, in electron transport experiments the addition of a glue layer can complicate the analysis, and in high energy density science experiments where shocks are launched into targets a glue layer can cause reverberations between the layers, again complicating the experiments. We have used flat and structured aluminium substrates to investigate the suitability of this technique for multi-layer target manufacture.

It is important to have good uniformity of electroplated foils when manufacturing them for laser experiments. Target foils are usually <100um thick. During electroplating nodule formation is common [5]. If nodules are allowed to form during the plating process foil thickness varies across the surface. This result will not meet target criterion. Inconsistencies in foil thickness can have an impact on the laser experiments for which they are produced; therefore one area of focus is to reduce the

IOP Conf. Series: Journal of Physics: Conf. Series **1079** (2018) 012016 doi:10.1088/1742-6596/1079/1/012016

number and height of these nodules as successful, uniform plating across flat and structured surfaces also offers the potential reduction of components and assembly processes in target manufacture.

Preliminary work carried out saw copper electroplated directly on to flat aluminium substrates without any prior treatment, other than cleaning the aluminium substrates. This work was done to establish valuable areas of investigation. Figure 1 shows the formation of islands on the aluminium substrate. Figure 2 shows the size variation of the copper islands. These plating results show nodules of up to 154um proud of the plated surface on a 240um thick aluminium foil.

There are many factors that can affect the quality of a plated surface, in this case five were chosen; effect of seed layer, pH of plating solution, solution mixing, current applied, and solution temperature. Copper was electroplated onto aluminium foils (dimensions: 0.24x10x40mm). Copper sulphate solution was made in the laboratory from solid copper sulphate crystals, 1 moldm^{-3} solution was made for each test. 240um thick and 99.0% purity aluminium sheets were cut into 5x1cm strips. All aluminium samples were cleaned with 10% sodium hydroxide solution to remove surface contaminants, following cleaning copper seed layers were deposited using plasma sputter coating on most aluminium substrates. When cleaned and coated, prepared aluminium strips were kept under nitrogen to prevent oxide formation on the surface. Native oxide layers are common and inevitable on aluminium and copper metals exposed to oxygen. [4, 5]

2.1 Seed Layer Effects
An aluminium test strip was plated with no seed layer in 1mol dm^{-3} $CuSO_4$ solution at 10mA to establish the baseline plating parameters. This test sample did show some copper deposition; however the results gave nodules or islands of copper rather than foils plated over the entire aluminium surface. Nodule formation is investigated in some detail in literature by Guo [6], [7]. Copper had not plated across the whole surface with the edges preferentially plated. Where plating had occurred copper nodules ran along furrows present in the aluminium foil. Results from this run achieved no usable foil or flat area. Nodules formed varied greatly in diameter and height. Figure 2 shows the diameter ranges observed. The modules characterised varied in height, from 3um-154um.

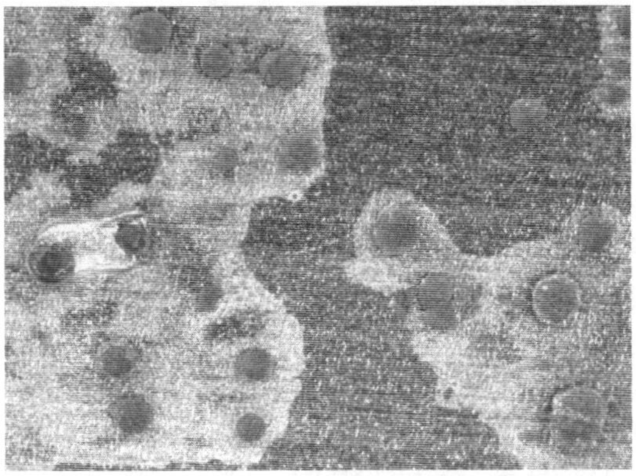

Figure 1: Seedless copper plating on aluminium substrate, 10mA applied, 2hours, 1mo dm^{-3} $CuSO_4$. 2.5x MAG.

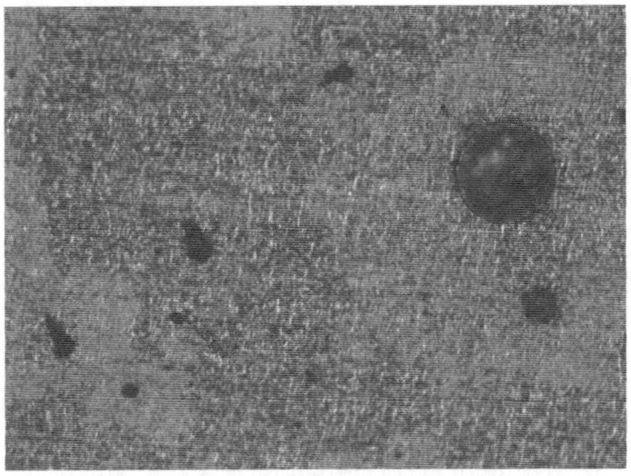

Figure 2: Close up of two nodules formed on aluminium substrate. Larger nodule has a diameter of 146um, smaller diameter 17um

A second test piece was coating with a seed layer of copper as described above. Seed layers deposited were 240nm on side A and 180nm on side B. It is clear that this layer greatly enhances the deposition of the copper during the electroplating process. The copper applied showed a marked improvement in uniformity and quality compared to results of plating cleaned aluminium. Figure 3

IOP Conf. Series: Journal of Physics: Conf. Series **1079** (2018) 012016 doi:10.1088/1742-6596/1079/1/012016

shows the plating attempt made with a Cu seed layer. Figure 3 also shows the variation in quality across the surface with areas of discoloured and bright copper. Although, the resultant foil is poor in copper quality, the image does show increased foil plating success across the substrate compared to attempts made without seed layers.

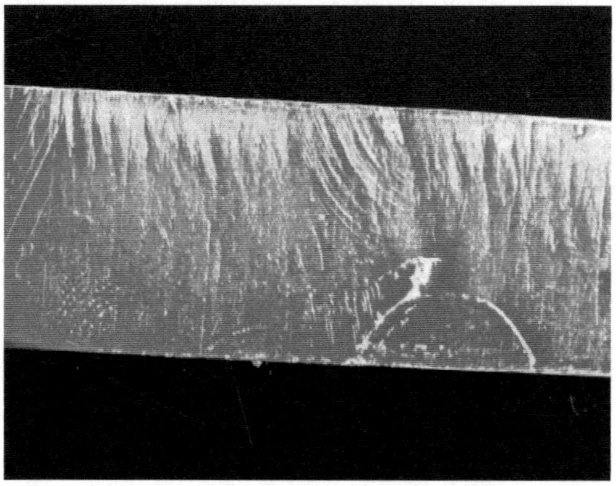

Figure 3: Copper plating with seed layer, 10mA applied, 2hours plating time, 1mol dm^{-3} $CuSO_4$ solution concentration.

2.2 Effect of Solution Temperature

The aim of investigating solution temperature was to assess if the thickness of the copper foil plated differed across a temperature range. Three temperatures were investigated; 10°C, 18°C and 40°C. Room temperature of the lab used is 18 °C. Plating attempts made at 40°C delaminated to some degree.

It was evident that there was some plating success; delamination of the plated foil was deemed unsuccessful. Although this information is useful, this investigation aimed to achieve multi-layered, glue-free targets. Side A (240nm seed layer) delaminated more readily and in larger pieces than side B (180nm seed layer). At 10°C there was little to no change to the foil thickness compared to foils produced at 18°C. The temperature difference may have been too small to cause any change in plating result.

2.3 Current Applied to the Sample

The current applied to the aluminium strips was 5, 10, or 15mA. All samples were processed in a standard set up of 1mol dm^{-3} $CuSO_4$ solution for 2 hours, without continuous mixing, at 18°C. Where 5mA was applied poor results were obtained. In these instances, the copper deposited only as nodules that ranged from 25um-110um in height. The nodules post-plating appeared bright copper coloured. It is unclear the reason for the blue colour change on the aluminium strip observed in figure 4. EDX was carried out to assess the elements identified; only copper and aluminium were identified with no other significant elemental peaks. Figure 4 shows one of the results of this plating attempt. No overall thickness measurements were taken for these samples as there were no areas where usable foil had been plated to obtain consistent measurements.

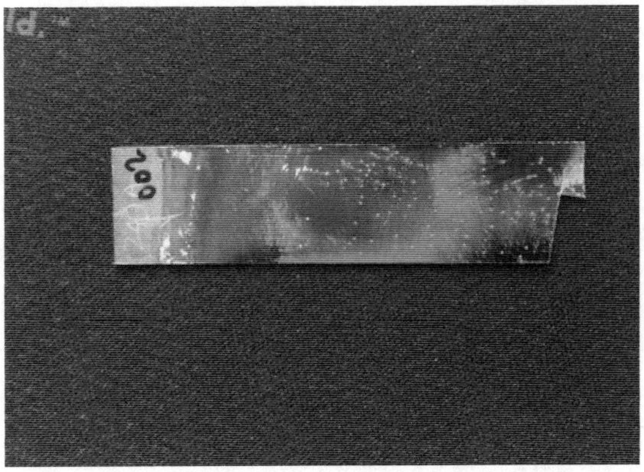

Figure 4: Image of 5mA no mixing, nodules can be seen across the surface. No foil success.

Where 10mA was applied again poor quality foils were achieved, this is shown in figure 4. Copper plating achieved on side A (240nm seed) gave thicknesses of 18-20um, side B gave foil thicknesses of only 6-8um. It is unclear the reason of thickness variation on sides A and B. The thicknesses vary throughout the investigation and there is no obvious preferred seed layer thickness at this stage. The roughness (Ra) was measured using white light interferometry to be 3.8um on side A and 1.8um on side B. Discolouration on the surface was observed along the width of the sample, following the orientation of the grooves found in the aluminium pre-processing.

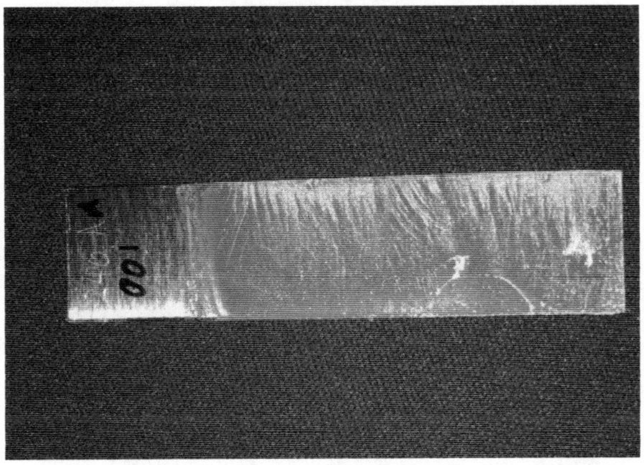

Figure 5: 10mA applied, solution was not continuously mixed. Copper thickness varies greatly across entire surface.

Where 15mA was applied foils of copper were achieved. These samples gave thicknesses of 0-28.8um on side A. Side B was uneven giving coating of 0um-17um. The thickness of copper across the surface was arbitrary. Both sides showed areas where plating had not occurred. Although copper had plated onto the aluminium surface, large areas of thick, nodular, dull copper and areas of thinner, flatter, brighter copper were observed. The thicker areas appeared darker in colour, much like the foils achieved at 10mA. The thickness range across the aluminium surface does not meet with the aim of the investigation. Results obtained suggest these parameters should not be used for the purpose of producing thin copper foils.

EDX analysis was carried out on the discoloured samples, significant oxygen peaks were observed suggesting the copper had oxidised. Figures 6 and 7 show images taken with a scanning electron microscope (SEM) of samples plated at 5mA (fig. 4) and 10mA (fig. 5). Although at 10mA plating had showed greater success than at 5mA, the surface roughness observed implies a greater copper surface area exposed to air, thus increasing the probability of oxide formation.

Figure 6: SEM (Pemtron PS230) Image of copper nodule formed on aluminium strip. Plating conditions: 18°C, 1mol dm^{-3} $CuSO_4$, not stirred throughout plating time, 5mA applied.

Figure 7: SEM (Pemtron PS230) Image of copper surface formed on aluminium strip. Plating conditions: 18°C, 1mol dm^{-3} $CuSO_4$, not stirred throughout plating time, 10mA applied.

2.4 Continuously Mixing the Solution

Initial tests carried out without mixing the plating solution gave inconsistent foils and nodules on the plated areas of the substrate. For these tests all other factors remained constant. Current applied was 5, 10, or 15mA in a 1mol dm^{-3} $CuSO_4$ solution, for 2 hours, at 18°C.

All test strips showed a marked improvement in uniformity of thickness across the surface when the solutions where continuously mixed. Dull areas remain, and this is where the copper was rougher compared to the appearance of brighter foil sections, foil roughness detailed in section 2.6. Thicknesses achieved can be seen in table 1. These foils adhered well to the aluminium surface making them useful for multilayered target manufacture.

Current applied (mA)	Solution Mixed - Thickness side A (um)	Solution Mixed - Thickness side B (um)	Not Mixed Thickness side A (um)	Not Mixed Thickness side B (um)
5	5	14-15	N/A too many nodules	N/A too many nodules
10	20-21	7-8	18-20	6-8
15	9-10	10	0-28	0-17

Table 1: Copper foil thickness using white light interferometry.

2.5 Variation of the pH of Solution

To reduce the pH of the solution sulphuric acid (10ml H_2SO_4 at 50%) was added to the copper sulphate. All other conditions remained the constant as in the previous tests the current applied was 5, 10, or 15mA in a 1mol dm^{-3} $CuSO_4$ solution, for 2 hours, at 18°C . Initial plating attempts made without solution mixing produced nodular foils and therefore due to the foil success achieved by continuous mixing as detailed in section 2.4 this factor was kept constant.

The resulting foils across all currents applied were flatter, appeared bright copper in colour and plated more evenly over the entire substrate. However these foils were easily removed from the

IOP Conf. Series: Journal of Physics: Conf. Series **1079** (2018) 012016 doi:10.1088/1742-6596/1079/1/012016

aluminium substrate by peeling. Though useful, our aim was to achieve copper suitably adhered to the substrate to produce layered targets. Figure 8 shows the flatness achieved at 5mA with sulphuric acid. It was observed that 5mA give the best quality foils and this is due to the slower plating rate at this current density. Figure 9 shows the comparison of foils with and without sulphuric acid added to the solution. Table 2 show the variation in thickness achieved under these conditions. At this stage, the variation in results implies difficulty in estimating foil thicknesses. Reducing the pH does give flatter, brighter copper foils. Further investigation is needed to identify the parameters to be set to improve repeatability and achieve reliable foil thicknesses.

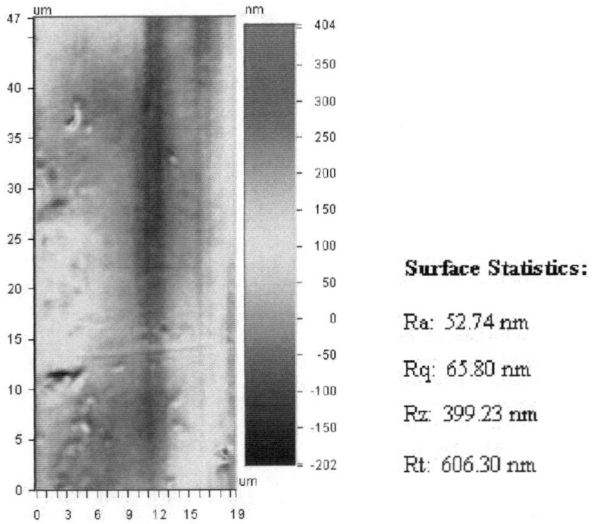

Figure 8: white light interferometry scan, 5mA 2 hours sulphuric acid. Roughness of 53nm achieved

Figure 9: comparison of plating results with and without sulphuric acid

Current applied (mA)	Copper thickness (um) $CuSO_4$ without H_2SO_4 Side A	Copper thickness (um) $CuSO_4$ without H_2SO_4 Side B	Copper thickness (um) $CuSO_4$ with H_2SO_4 Side A	Copper thickness (um) $CuSO_4$ with H_2SO_4 Side B
5	N/A too many nodules	N/A too many nodules	2.5	2.5
10	18-20	6-8	20	6

IOP Conf. Series: Journal of Physics: Conf. Series **1079** (2018) 012016 doi:10.1088/1742-6596/1079/1/012016

15	0-28	0-17	5	15

Table 2: Copper foil thickness range for solutions with and without sulphuric acid

2.6 Characterisation of Foil Roughness
Roughness of the foils deposited was also characterised as this is an important factor for high power laser targets where there is often a tolerance on surface roughness for the foils. It was seen that the roughness improved (Ra reduced) with continuous mixing and with the addition of sulphuric acid. The foils deposited at 5mA current were bright, with a near mirror finish. The foils were characterised using a white light interferometer.

Current Applied mA	Roughness (Ra)(nm) $CuSO_4$ No H_2SO_4 No mixing Side A	Roughness (Ra) (nm) $CuSO_4$ No H_2SO_4 No mixing Side B	Roughness (Ra) (nm) $CuSO_4$ No H_2SO_4 Mixed Side A	Roughness (Ra) (nm) $CuSO_4$ No H_2SO_4 Mixed Side B	Roughness (Ra) (nm) $CuSO_4$ With H_2SO_4 Mixed Side A	Roughness (Ra) (nm) $CuSO_4$ With H_2SO_4 Mixed Side B
5	N/A too many nodules	N/A too many nodules	210	180	N/A delamination	53
10	3800	1800	1000	1240	240	240
15	N/A sample too dark to characterise.	N/A sample too dark to characterise.	790	760	400	470

Table 3: Roughness (Ra) scans using white light interferometry

2.7 Processing Conditions for Optimal Laser Targets
There are many factors that can affect the quality, thickness and usability of plated metals. The five factors investigated here in this preliminary research have shown some improvement in foil success. It can be seen that if layered copper on aluminium targets are required, the best results would be achieved with foils that are electroplated with continuous mixing of the solution and the addition of sulphuric acid, with a seed layer (180nm) and at 5mA. This slow plating rate gave good uniformity and improved surface roughness allowing for adhesive-free, layered metal targets.

From this preliminary investigation there are clear opportunities to improve the quality of the plated foils. Work is planned to investigate the impact of the seed layer thickness. This is important as seed layers also give the opportunity to plate onto non-conductive materials, thus further opening up the option of adhesive-free layered targets. Polishing the aluminium substrate to remove the furrows present in the foils is also an area of further investigation to consider.
We hope to develop capabilities to plate metals on to machined structures, which could further reduce the components and assembly in target manufacture.

3. Ultrathin polymer film
We investigated films that are 90nm or less, that are used as a high-power laser target suspended (in most cases) over a hole to produce a membrane target and in this report ultrathin polymer films are made from polyvinyl formal resin. The resin is formed from raw material purchased in solid powder form that is mixed with a solvent to allow coating to take place using dip coating and spin coating methods. Solvent selection included establishing appropriate drying times to eliminate the solvent used in the manufacture process whilst still producing a useful film.

The polymer used for these films has been commercially known as Formvar, (a poly(vinyl) formal) and has traditionally used with TEM grids in microscopy. The original driving force of this work was to produce support films for indirect drive inertial confinement fusion experiments. The support films would sit within the gold hohlraum and support the spherical fuel shell. To satisfy the requirements of this target the film needs to be thin and strong enough to hold the shell in place without imprinting upon the target when compressed [8]. Thin films have also been used in high power laser experiments looking at efficient proton acceleration via a hybrid scheme of radiation pressure-sheath acceleration in an ultrathin foil irradiated by a linearly polarised laser pulse [9]. Metrics to assess film suitability included looking at film thickness, uniformity across an area, removal from the substrate, and ability to be used as a supporting film.

3.1 Assessing the roughness of the substrate

A profilometer (KLA Tencor Alpha-step IQ) was used to analyse the surface roughness of glass slides. In some cases where the characterisation was not clear due to the resolution limitations of the system, atomic force microscopy was used to assess the roughness. Glass slides that were the smoothest were picked as the substrates in the film study.

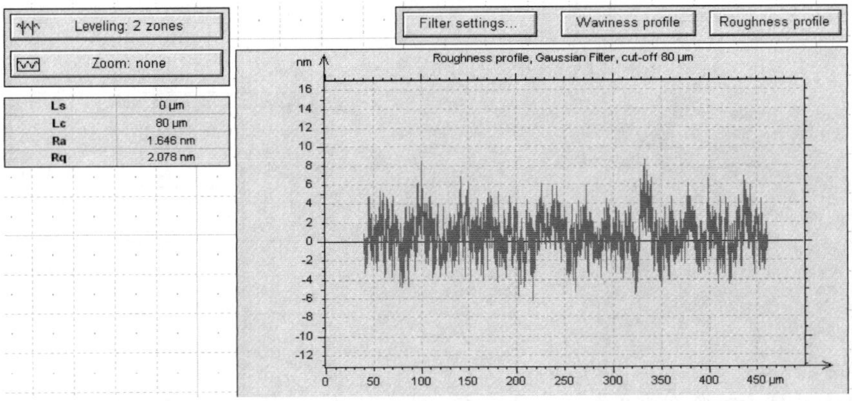

Figure 10. Roughness scan of a microscope slide used in the investigation (Ra = 1.646 nanometers). Scanned on profilometer (KLA Tencor Alpha-step IQ).

3.2 Coating Processes

Preliminary work was carried out to establish a procedure to obtain the thinnest films and both spin coating and dip coating techniques were investigated. Spin coating is a well-established coating technique that employs centrifugal forces to spread a liquid over a spinning substrate. Using spin coating it was found that to produce the best films applying the mixture to the substrate after ramp up (acceleration of the substrate) was ideal. If solution was deposited before the substrate was set to spin the resultant films would exhibit a waviness across the surface. Spin coating did produce thin films; however, they were all above the 90nm and it was not possible to achieve a thinner coating that was uniform across the surface.

Dip coating was the preferred method of film formation because it produced the thinnest films in the preliminary investigation. Films are produced by vertically dipping a flat substrate, in our case glass microscope slides, into the solution and lifting them out at a set rate. The dip rate greatly affects the film thickness, as does the solution concentration. The faster the substrate is pulled out of solution the thicker the resultant film. Once removed from the solvent-resin mixture the slide is left to dry. The solvent quickly evaporates from the slide, leaving behind a film of Formvar. Using this technique, films as thin as 10nm were achievable.

The greatest difficulty encountered during this investigation was removing the film from the glass substrate because as thinner films were achieved they became increasingly difficult to remove from the substrates (this was also apparent with other substrates such as polished silicon wafers). Typical float-off techniques were used, without a release layer, to remove films from the substrates and to help improve ultrathin film release substrates were cleaned using oxygen plasma cleaning. The films were then lifted from the water onto a support frame.

IOP Conf. Series: Journal of Physics: Conf. Series **1079** (2018) 012016 doi:10.1088/1742-6596/1079/1/012016

3.3 Characterisation

To characterise the films from both spin and dip coating a surface profilometer is used. Where films are too thin for to resolve the thickness using a profilometer atomic force microscopy is used. The films on the substrate are scored using a scalpel blade to give a step height that can be measured. In other cases, witness slides are masked to give a step for measurement.

Following initial thickness characterisation a second check of the film thickness is made as the films are floated off from the glass substrate onto water and then back on to a clean flat substrate. (This can be another glass microscope slide or polished silicon wafer.) The benefit of this second characterisation step is the substrate will not be damaged from the scoring process giving increased confidence in film thickness characterisation. Film uniformity was also looked at as a factor of success and this was characterized by looking at the thickness across a number of scans as shown in figure 11.

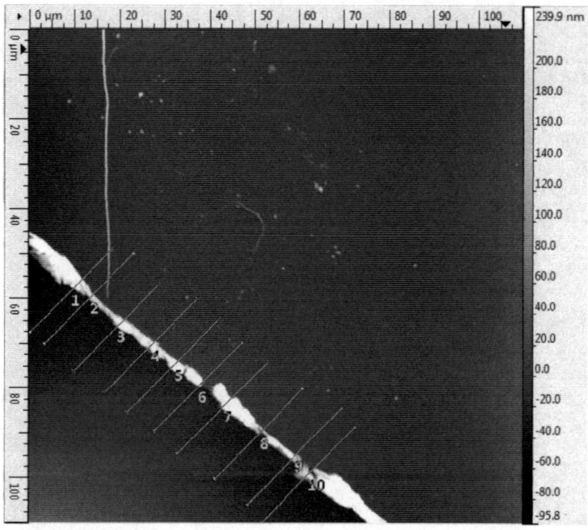

2015-11-06 0.2% Formvar mounted on Si 142	
	Height (nm)
Profile 1	35.9
Profile 2	34.4
Profile 3	30.0
Profile 4	29.4
Profile 5	28.2
Profile 6	24.3
Profile 7	29.7
Profile 8	32.9
Profile 9	28.6
Profile 10	38.6
Average	31.2

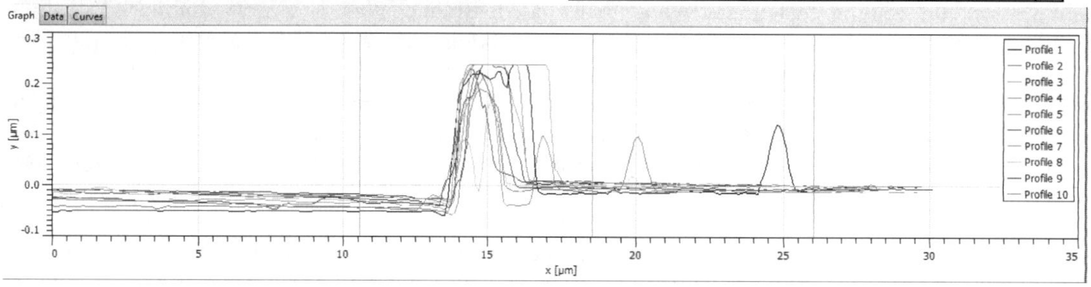

Figure 11. Lineouts across the edge of a formvar thin film using AFM

Ultrathin Formvar films were successful using dip coating procedures and the thinnest films achievable were 10nm thick. Further work is to be carried out to assess the strength of these films. In addition, work is required to improve the quality of the target frames as roughness of the edges on the mounts used in holding thin films could rupture the films.

References

[1] Serra S 2012 *Central Laser Facility Annual Report 2010-2011* 58

[2] Aasia Hughes 2016 *Central Laser Facility Annual Report 2014-2015* 52

[3] Sam Serra 2011 *CLF Annual Report 2009-2010* 45

[4] Jeurgens L, Sloof W, Tichelaar F, Borsboom C, Mittemeijer E. 1999 *J. Appl. Surf. Sci.* **144** 11-15

[5] Skarlinski M, Quesnel D 2015 *J. Appl. Phys.* **118**, 235306

[6] Guo L 2010 *Electrochinica Acta.* **55** 4086-4091

[7] Guo L, Oskam G, Radisic A, Hoffmann P, Searson P 2001 *J. Phys. D: Appl. Phys.* **44** 443001

[8] Nagel, S. R. 2015 Lawrence Livermore National Laboratory *Effect of mounting membrane on the shape of ICF implosions.* LLNL

[9] Higginson A, 2018 *Nature Communications* **9**, 724 2018

IOP Conf. Series: Journal of Physics: Conf. Series **1079** (2018) 012017 doi:10.1088/1742-6596/1079/1/012017

Surface roughness of NaCl coating used as release layers in thin film production

P Ariyathilaka[1], D Haddock[2], C Spindloe[1 2] and M K Tolley[1 2]

[1]Scitech Precision Limited, Rutherford Appleton Laboratory Harwell Campus, Chilton, Didcot, OX11 0QX, UK.
[2]Central Laser facility, Rutherford Appleton Laboratory Harwell Campus, Chilton, Didcot, OX11 0QX, UK.

E-mail: Pawala.Ariyathilaka@scitechprecision.com

Abstract. Surface roughness scans of NaCl (salt) coatings on silicon wafers were carried out using an Atomic Force Microscope. Metrology was carried out on single layer salt and also salt coated with a thin aluminium layer which is a representative high power laser target. All were characterised straight from vacuum and again after being exposed to the atmosphere for a few days. Results suggest that the surface roughness increased drastically after being left in atmospheric conditions for few days. The roughness increased by almost five time for an exposed salt coating. For a salt coating with an aluminium layer the roughness only increased slightly (~1nm). This suggests that it's important to coat materials for the laser target on to salt release layers almost immediately from vacuum. This work enables production processes and quality control for high power laser targets to be defined and an understanding of possible pinhole production in thin film coatings.

1. Introduction

The use of salt as a release layer for thin film coatings to prepare targets for high power laser experiments has been used for many years [1]. In addition a soap solution, or other salt material can also be used, with each type or release mechanism having its benefits. The use of salt coated by vacuum deposition is desirable as the release layer thickness can be controlled precisely using the thin film crystal monitors in vacuum coating plants. A release layer is essential when fabricating thin film targets, as it is designed to stop the metal coating from adhering to the supporting substrate material. These slides can be lowered in to de-ionized water to lift the metal foil by dissolving the salt. The water tension keeps the metal foil suspended on top, which allows to pick up even a very delicate foil directly with the target mount.

Whilst work has been carried out to understand the release nature of the salt coating, the behaviour over time of this and its effect on the laser target is to be studied. It is important to have a clear idea of the origin of the surface roughness of this layer as any metal coated on the surface follows the same form. To provide consistent thin film targets, that are pinhole free and that survive at ultra-low thickness with a good surface finish, the surface of the release layer must be studied.

2. Experimental set up

In this report, two main conditions were investigated. The first parameter studied was the surface roughness study of a salt release layer straight from the vacuum chamber. The second parameter was the study of salt surface after being left out in the atmosphere for a week. The humidity of the class

10000 clean room is around 40 ± 2 percent. All the coatings were carried out using Physical Vapour Deposition (PVD) using a thermal evaporation source with a limited source size. The test samples used in this experiment were pieces of Silicon (Si) wafers with roughness of <1nm [2]. The coatings were repeated three times. The average values of the three are quoted in this paper. The Si wafers were plasma cleaned using an Oxygen plasma prior to any coating. A salt coating of 100nm was coated on to two sample Si wafer pieces. This method involves heating the material (salt in this case) until it sublimes. The salt vapor is then allowed to deposit on the Si wafer producing a coating [3].

For an effective PVD process and a high quality coating, the Si wafer needs to be in a good vacuum to have little residual water content on the substrate as well as to give a good mean free path to the sublimed salt. The chamber pressure for the salt coating was 3.4×10^{-6} mbar. Two Si wafer pieces were coated with salt. Immediately after coating, one of the Si pieces was removed and placed under an Atomic Force Microscope (AFM) to be characterised. The second sample was pumped back down to a vacuum of 2.5×10^{-6} mbar and coated with 100nm of aluminium (Al). During this process of removing the first sample the chamber was vented to dry nitrogen to avoid contamination and immediately pumped back down. A load lock, although desirable was not available in this case. The second sample was measured after it was coated with aluminium and then both measurements were repeated after being left at atmospheric conditions for a few days.

3. Data Collection

Figure 1 below shows the way data collection took place. Both Si samples were scanned using an AFM to collect the surface roughness data. The AFM used in this experiment was a Veeco diCaliber with a resolution of around <0.6nm in the z direction.

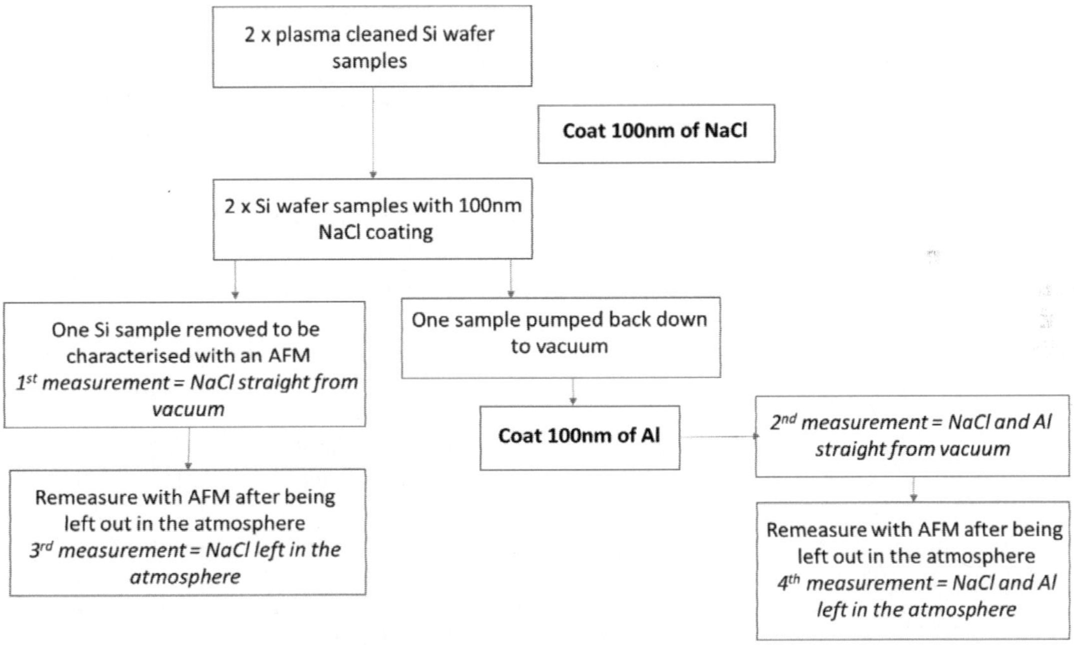

Figure 1. Image shows the data collection carried out in this project.

AFMs can operate in scanning mode or in tapping mode. In this experiment tapping mode was used. The AFM operates by scanning a fine tip attached to the end of an oscillating cantilever across the sample [4], which gives the morphology of the surface. The cantilever oscillates close to its resonance frequency, and lightly taps the surface as the sample is scanned. An incoming laser beam reflects off the cantilever, on to a position sensitive photodetector which contains a four-segmented photodetector [5] as shown in figure 2.

IOP Conf. Series: Journal of Physics: Conf. Series **1079** (2018) 012017 doi:10.1088/1742-6596/1079/1/012017

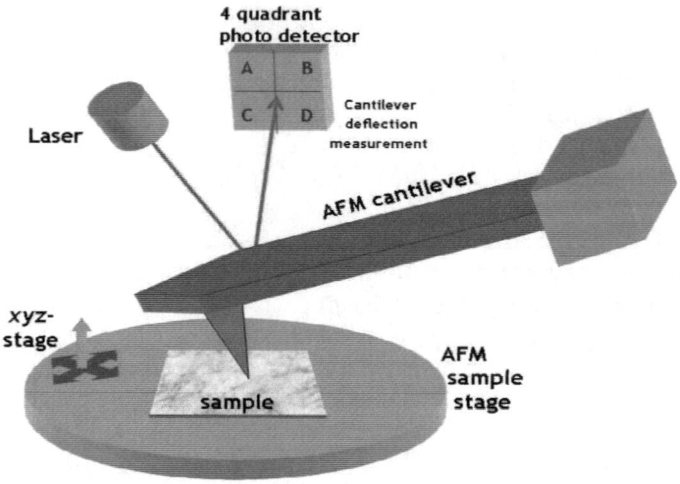

Figure 2. An atomic force microscope in action [6].

Surface characteristics can be described on different length scales, roughness for high spatial frequencies and waviness for variations on larger scale. In this report, surface roughness was studied further which is the finest irregularities of a surface that arises from a production process of a material. The AFM tip used for the measurements in this report has a radius of curvature of 20nm. The AFM settings were as below. The range for the scan was 5.1969um with 1024 pixels per line and a scan rate of 0.1Hz.

4. Data analysis software – Gwyddion

The AFM data generated needs to be analyzed to give it clarity. Gwyddion is ideal for flattening and filtering AFM images to look at the surface in more detail. All the images in this report went through the following flattening process [7].

- Plane level - Which does a mean plane subtraction for the full set of data. The plane is computed from all the image points.
- Align using rows - This shifts the lines in the x axis such that the median of difference between vertical neighboring pixels is zero. This keeps rather large features.
- Remove scars - These are parts of the scan that is affected by common scanning errors. This function corrects for this.

With the AFM tip being 20nm in radius, a Fourier filter was applied to remove the very high frequencies with a cut off wavelength of 1/20nm. Once the image is leveled and filtered, the roughness is measured using the roughness tool. An image of an example 1D roughness scan is shown in figure 3. To measure the roughness, a cut off value needs to be set. The cut off in Gwydion is in terms of Nyquist frequency, this was set to 0.003 in this report, with a pixel size of ~5nm this equals to a cut off wavelength of ~3um. It should be mentioned that the roughness measurements here are not done exactly to ISO standards. The same cut off is chosen for all the data for a direct comparison.

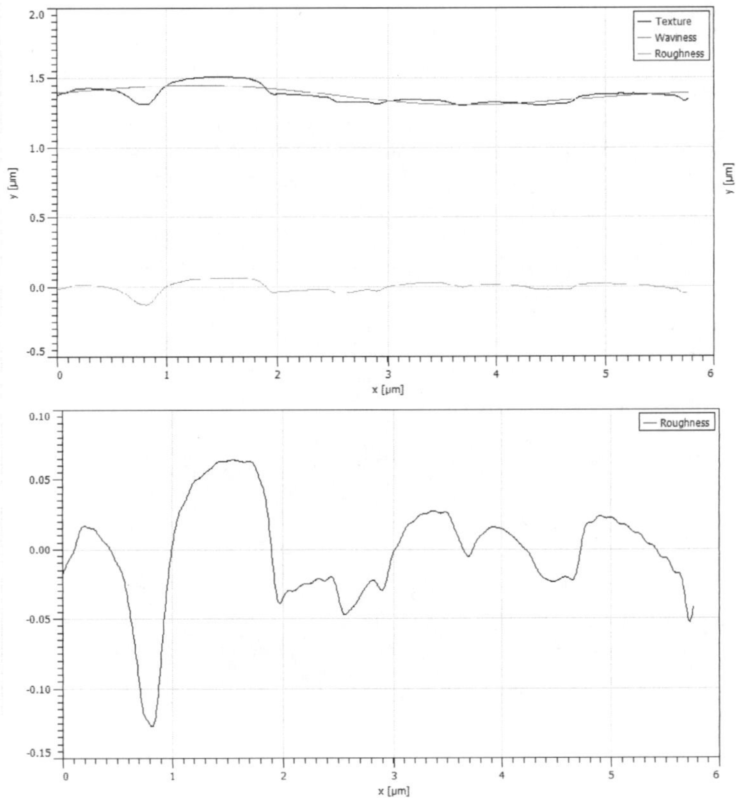

Figure 3. Shows the roughness measurements in Gwyddion. For this particular 1D roughness scan, the roughness was 29.5nm. This figure is for a scan of a NaCl sample exposed to atmosphere.

Within Gwyddion, 1D roughness is defined as 'Arithmetic mean deviation. The average deviation of all points roughness profile from a mean line over the evaluation length [8]'. This evaluation line is drawn through the filtered image in five different places as shown in figure 4.

IOP Conf. Series: Journal of Physics: Conf. Series **1079** (2018) 012017 doi:10.1088/1742-6596/1079/1/012017

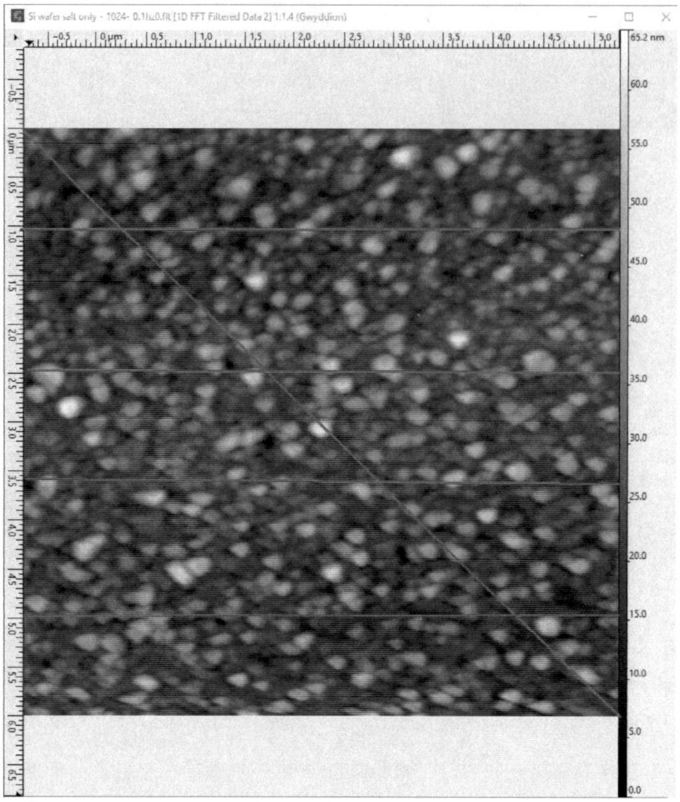

Figure 4. Image above shows where different 1D roughness lines were taken. The roughness lines are shown in red. The average roughness values quoted in the paper are the averages of the above lines, averaged over the three repeated samples. Also note that the 1D roughness are drawn separately when measuring the roughness. The above image is only drawn to show the 1D roughness lines.

5. Discussion

The roughness values are shown in figure 5. The average roughness for a salt coated Si piece from vacuum was 7.7nm which increased to an average of 38.1nm when left out in the atmosphere which was an increase by almost 500%. This can be attributed to the salt grains being crystalized once in contact with the moisture in air. The salt and Al coated Si wafer sample from vacuum gave a roughness value of 3.5nm, which increased to 4.7nm once left in atmosphere for a few days, a change of approximately 1nm. It is noticeable that a salt coating left out in has a much higher roughness variation over time than that of the coating of salt and aluminium. A possible explanation could be that the aluminium coating has sealed the salt completely from the atmosphere, or limits the moisture penetration, limiting the size of the crystals. The difference in the two example salt coatings are shown in figure 6.

IOP Conf. Series: Journal of Physics: Conf. Series **1079** (2018) 012017 doi:10.1088/1742-6596/1079/1/012017

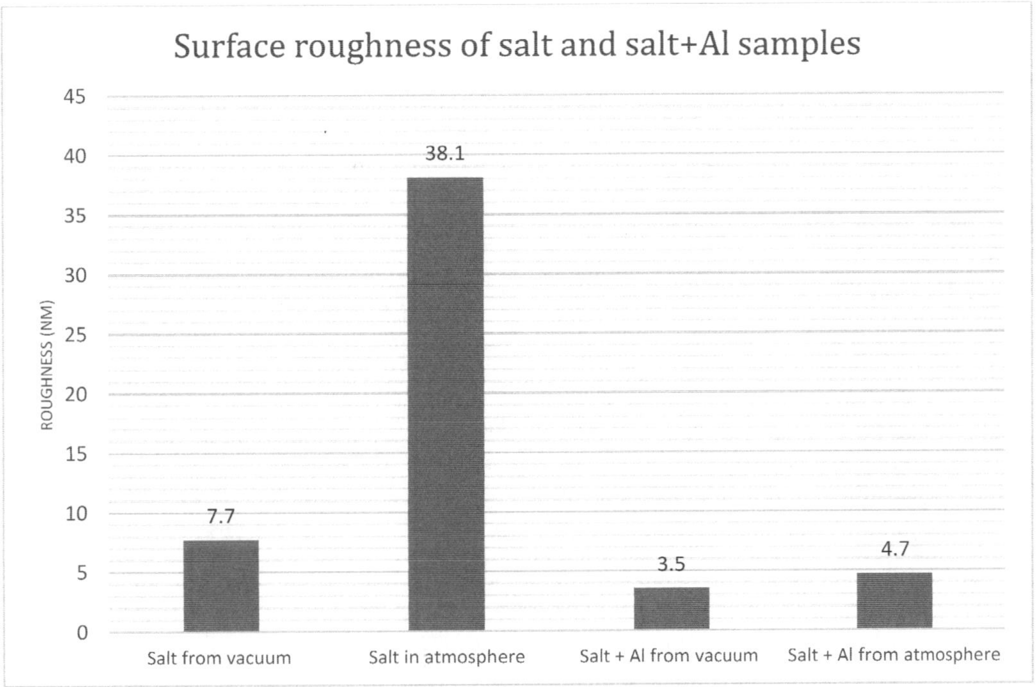

Figure 5. Shows the roughness of the samples

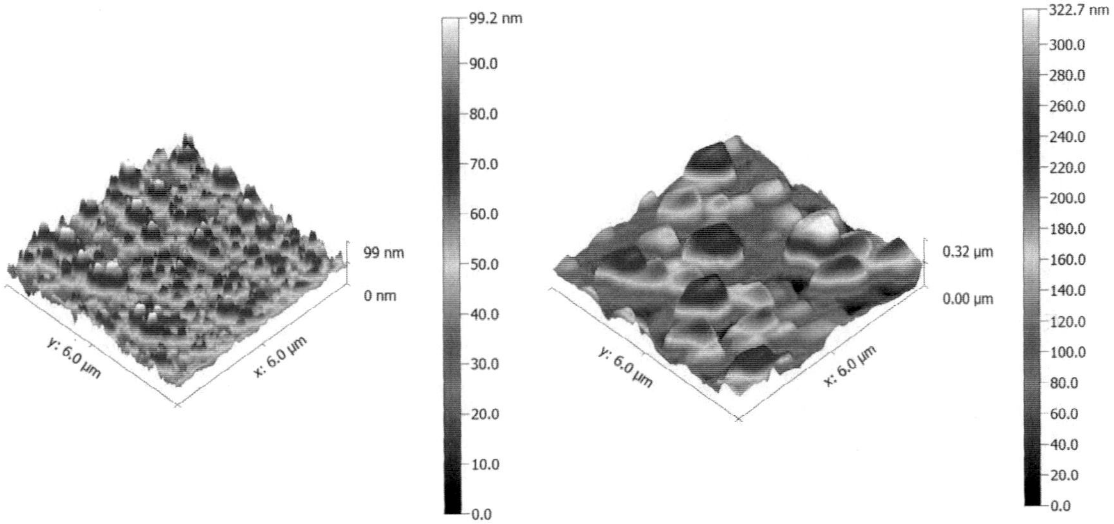

Figure 6. Image showing a comparison of a coating straight from the vacuum (left) vs being out in the atmosphere (right).

After the surface roughness scans were complete, the test Si wafers samples which had salt and aluminium coatings were floated off using the standard method of dipping in de-ionised water. The purpose of this was to see whether there's any big discrepancies between the measurement on a wafer and a measurement that was carried out after processing the foil into a target geometry. The average roughness for a floated off foil (now single layer aluminium as the salt is dissolved) was 4.1nm, suggesting that the slightly larger roughness noticed on the salt and aluminium from atmosphere samples were only present on the salt later and hasn't influenced the floated off foil. An AFM scan was done on

IOP Conf. Series: Journal of Physics: Conf. Series **1079** (2018) 012017 doi:10.1088/1742-6596/1079/1/012017

both sides of the floated off film. The roughness on both sides show no significant difference. It is possible to infer that the morphology of the salt hasn't transferred completely to the coating when it is removed from the salt substrate.

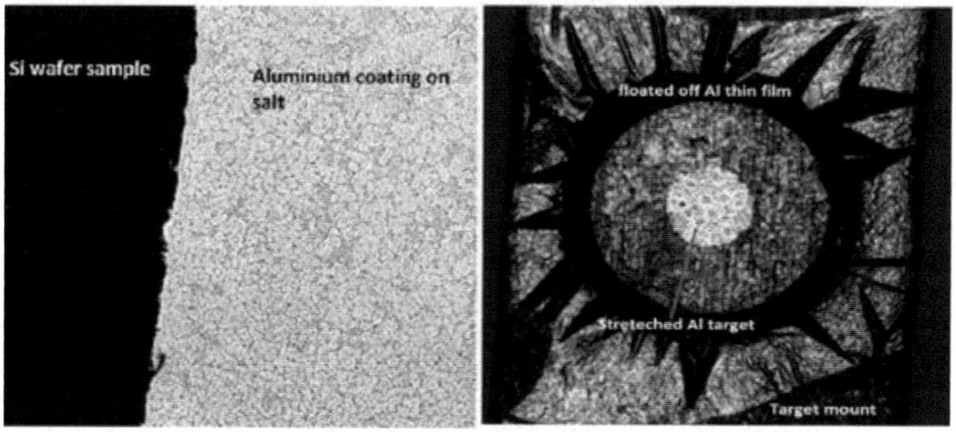

Figure 7. Photograph showing optical micrograph images of Al coating on Si wafer (left) and an assembled Al target on a Cu mount. (right)

6. Conclusion

This report looked at the surface roughness of salt (NaCl) coatings that are used as release layers in laser target manufacture. Two samples were looked at in detail with each coating being repeated three times. One sample had a coating of 100nm salt only and the second sample had 100nm salt followed by 100nm of aluminium. The roughness was measured with an AFM for all the samples. For both samples, one measurement was done straight from the vacuum chamber and the second was once the samples were left out in the atmosphere for a week.

The results found show that the roughness of both the samples increased after they were left out in the atmosphere with the single layer salt coating showing the largest and most significant roughness increase. It is concluded that the reason for this is the crystallization of the salt grains due to the moisture in the air. The salt coating with an aluminium top coating showed reduced increase in roughness, and this is suggested to be due to the barrier layer that is introduced between the salt and the atmosphere preventing some crystallization.

It is also noted that when a target foil (the aluminium layer) is removed from the substrate (salt) and suspended over a target support, a similar flatness is observed as for the unexposed target foil on a substrate, even after it has been exposed to a few days out in the atmosphere, and whilst that is encouraging for target foils, for filters and applications that require pinhole free foils, the stress on the thin layer in being on a rough surface and then flattened out again could cause small tears and pinholes in the foil. In further studies the same experiment could be carried out with varying salt thicknesses to determine whether this has any effect on the roughness of the target foil. To achieve a flat coating for a laser target, the time between the salt release layer coating and target material coating must be minimum. However exposing the release layer to atmosphere could be a way to develop some surface structure in a controlled manner.

References

[1] Wenjun Ma et al 2011 *Nucl. Instr. and Meth. A,* doi:10.1016/j.nima.2011.06.019
[2] C.Teichert et al Appl. Phys. Lett. 66 (18), 1 May 1995; doi: 10.1063/1.113978 Comparison of surface roughness of polished silicon wafers measured by light scattering topography, soft-x-ray scattering, and atomic-force microscopy.
[3] M Tolley and C Spindloe. *Laser-Plasma interactions and Applications*. Chapter 17. Microtargetry for high power lasers
[4] Veeco Dimension 3100 AFM user manual. Accessed via: https://www.nist.gov/sites/default/files/documents/cnst/nanofab/Veeco_Dimension3100_AFM_USERMANUAL_v1.pdf
[5] Wenjie Mai. *Fundamental theory of AFM.* Accessed via: http://www.nanoscience.gatech.edu/zlwang/research/afm.html
[6] Atomic force microscope image. https://simple.wikipedia.org/wiki/Atomic_force_microscope
[7] P.Klapetek, D.Necas, C.Anderson. *Gwyddion user guide. One dimensional roughness*
[8] P.Klapetek, D.Necas, C.Anderson. *Gwyddion user guide. One dimensional roughness parameters*, chapter 4.

IOP Conf. Series: Journal of Physics: Conf. Series **1079** (2018) 012018 doi:10.1088/1742-6596/1079/1/012018

MEMS Fabrication of Silicon Microwire Targets.

Graham Arthur

Scitech Precision Limited, Rutherford Appleton Laboratory, Harwell Campus, Didcot, Oxfordshire, OX11 0QX, UK.

graham.arthur@scitechprecision.com

Abstract. Micro- and nano-structured surfaces can lead to increased absorption of incident laser pulses in high power laser experiments. Here we describe how MEMS (Micro-Electro-Mechanical System) manufacturing techniques (in this case, optical lithography, and deep silicon etching), have been used to create surface structured ("microwire") laser targets consisting of 15um tall, regular arrays of 2um or 3um wide silicon microcolumns for a recent experimental campaign using the Vulcan laser at Rutherford Appleton Laboratory (RAL)

1. Introduction.

Recent high power laser experiments such as those described in [1] and [2] have shown that micro- and nano-structured surfaces can lead to an increase in the absorption of the incident laser pulses. There are a number of ways such a surface may be manufactured but one of the best routes is to use MEMS-based fabrication techniques to create regular arrays of closed-packed micron-scale pillars (microwires), which behave in the desired way. Some alternative methods (e.g. laser heating or electroplating) are able to create microwire targets of the correct dimensions and microwire densities but with a more random placement of wires within the target areas, whereas the chosen route allows pillars, grating or similar structures to be defined over specific areas which are both highly uniform and scalable. Other MEMS processes including wet etching [3] have been used elsewhere to create silicon micro-needles of similar dimensions for other applications such as drug delivery, but these have typically been pointed needles rather than the close-packed, square-topped, vertical-sidewalled pillars specifically requested for the experiments at RAL.

This paper describes the basic processes of MEMS-based fabrication and continues by outlining the process steps required to manufacture these arrays which consist of 2um or 3um microwires each over an area measuring approximately 500um x 500um per target. The microwire arrays were successfully created and have been used in a recent experimental campaign [4].

2. MEMS-based manufacture.

The MEMS-based manufacture uses technologies originally developed from those found in the semiconductor manufacturing industry. Unless requested otherwise, the substrates for this type of target fabrication are usually silicon wafers which are typically 100mm in diameter, have a thickness of 300-500 microns and are flat from edge to edge on the micron scale. Crystal orientation is normally <100> unless specifically requested otherwise. For target manufacture, wafer doping type is not normally customer-specified. MEMS microfabrication techniques are based upon the three basic processes outlined below:

IOP Conf. Series: Journal of Physics: Conf. Series **1079** (2018) 012018 doi:10.1088/1742-6596/1079/1/012018

2.1. Deposition.
This includes sputter-coating, thermal evaporation, chemical vapour deposition, thermal oxide growth and spin/dip coating. Consequently, it is possible to deposit precisely controlled films of metals, dielectrics and polymers.

2.2. Etching.
This may be either dry etching or wet etching. In the former, a plasma containing highly reactive species (e.g. fluorine, chlorine) is generated which etches the substrate and/or previously deposited thin films through a protective masking layer. In the latter, the substrate and its coatings are immersed in a solution which is chosen to etch the required layer, whilst leaving the masking and other layers unaffected.

2.3. Pattern transfer.
This is the key process and, depending upon the dimensions required for the laser target, uses either optical, e-beam, or nanoimprint tools to pattern a thin film of material known as a resist which has previously been applied to the substrate by spin- or dip-coating. The exposed and developed resist film is then used as a protective stencil through which deposition or etching can take place, thereby transferring the pattern into the substrate and/or its coatings.

Optical lithography uses ultraviolet (uv) light passing through photomasks to define the patterns in an optical resist. Resolution is limited by the wavelength of the light and for laser target-making this is usually restricted to about 0.5-1 micron.

For the highest resolution features, electron beam lithography is used. Here an electron beam focused down to a nanometre-scale spot is driven across the surface of the resist to directly expose ("write") the pattern.

Nanoimprint lithography (NIL) uses a previously made stamp which is pressed into the resist layer. Application of heat or uv light retains the topography of the stamp in the resist thereby transferring the pattern. The stamp is also made using lithographic and etching processes. This method can achieve resolutions is the nanometre scale and is particularly attractive for mass production when compared to e-beam lithography.

These three basic processes are repeated as often as required, and in a pre-determined order until the final structures are created.

3. Target Specification.
Figure 1 shows the dimensional specification of the microwire target, whilst Figure 2 shows how this is realized as a practical laser target. The target mount is a 2mm x 7mm x 350um piece of silicon with the microwire array 2mm from one end. The trench around the target allows it to be easily separated from the silicon wafer as it is only secured with a small tab at the other end following fabrication and can be simply "snapped out" and bonded to a standard support structure.

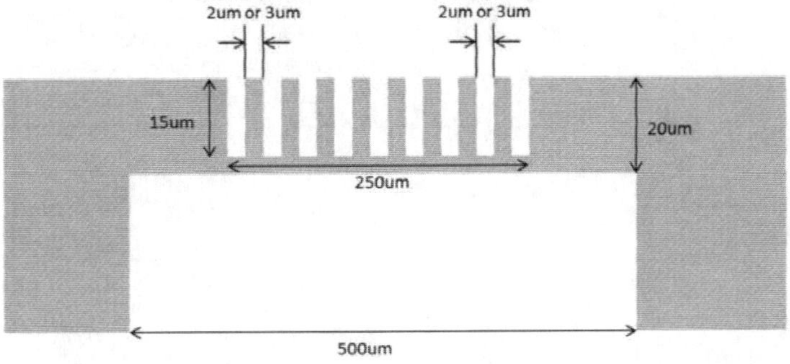

Figure 1. Basic target specifications.

IOP Conf. Series: Journal of Physics: Conf. Series **1079** (2018) 012018 doi:10.1088/1742-6596/1079/1/012018

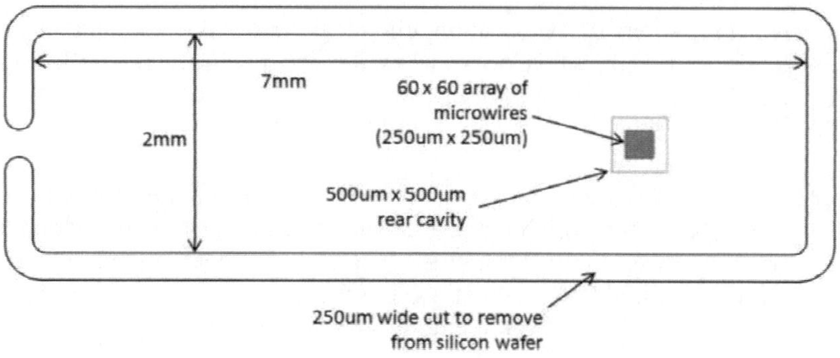

Figure 2. Microwire array as realised in a 2mm x 7mm silicon mount.

4. Target Fabrication.

Figure 3 shows the microfabrication steps, in cross-section, using a silicon-on-insulator (SOI) wafer as it is being processed. This wafer has a 20um thick silicon device layer and a 325um thick silicon handle (support) layer, separated by a 1um oxide layer, which acts as an etch-stop layer during the processing. Optical lithography using a Karl Süss MA-6/BA6 mask aligner was used throughout to define the structures which have a minimum feature size of 2um. Silicon etch was carried out using an STS deep reactive ion etch (DRIE) system and the oxide removal using an Oxford Instruments System 90 reactive ion etch (RIE) tool.

The process steps were as follows: Step (1) is the initial lithographic step which images the array of pillars into a 1.5um resist film on the device layer. Step (2) is the first deep silicon etch, creating the pillars in the device layer. Following a resist strip, step (4) is the next lithographic step which images the cavities and separating trenches into a 7um resist film on the handle wafer side. Step (5) is the second, deep silicon etch, from the underside, which stops at the oxide etch-stop layer. After another resist strip, step (7) uses a laser cutting tool to cut through the oxide and device layer, leaving each target array held in place by the small tab (see §3). Step (8) is a manual "snap-out" of the individual target arrays, as required. Step (9) is an oxide strip, using RIE, to remove the etch-stop layer prior to manual bonding of the target onto a standard support structure which is then placed in the target chamber.

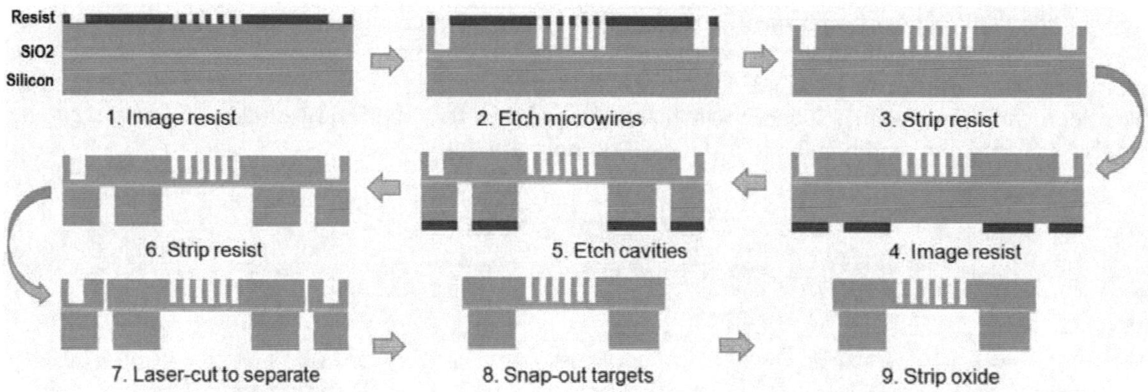

Figure 3. Basic MEMS-based process steps for the fabrication of the silicon micro-wire targets.

IOP Conf. Series: Journal of Physics: Conf. Series **1079** (2018) 012018 doi:10.1088/1742-6596/1079/1/012018

5. Results.

Following the process steps described above, 60 x 60 arrays of 2um and 3um microwires were successfully manufactured. Figures 4 to 6 show the microwire arrays during manufacture and inspection. The layout on the wafer meant that more than 100 target arrays were processed per wafer and, as only a fraction of the total available were required for this experimental campaign, high yield was not an issue and was not specifically measured. It does, however, demonstrate the potential for mass production if a similar type of target is required in the future.

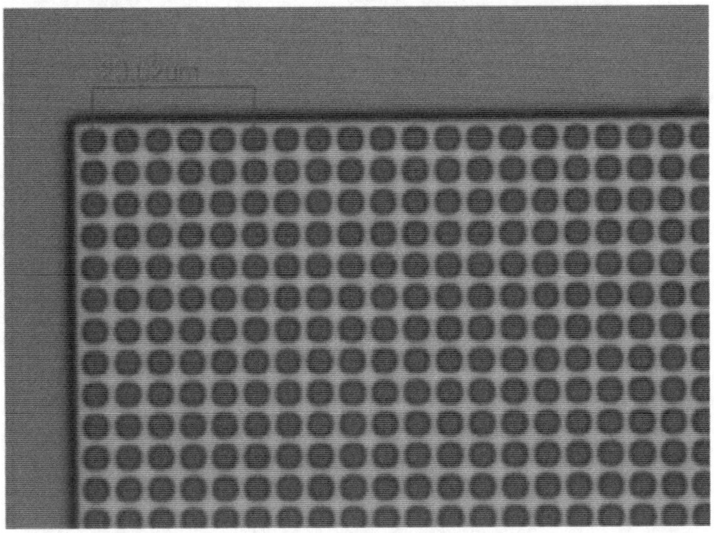

Figure 4. Optical image of the photoresist pattern of the 2um pillars before etching. This corresponds to Figure 3, Step 1.

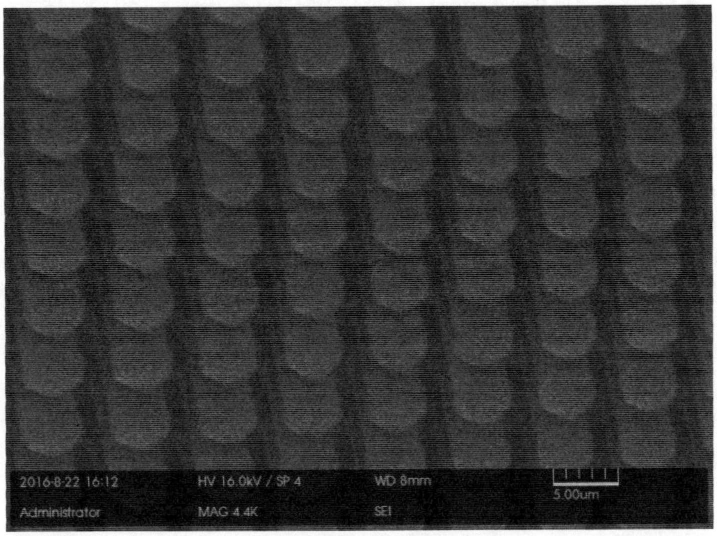

Figure 5. SEM image of 3um pillars after deep silicon etch viewed at 30°. This corresponds to Figure 3, Step 3.

IOP Conf. Series: Journal of Physics: Conf. Series **1079** (2018) 012018 doi:10.1088/1742-6596/1079/1/012018

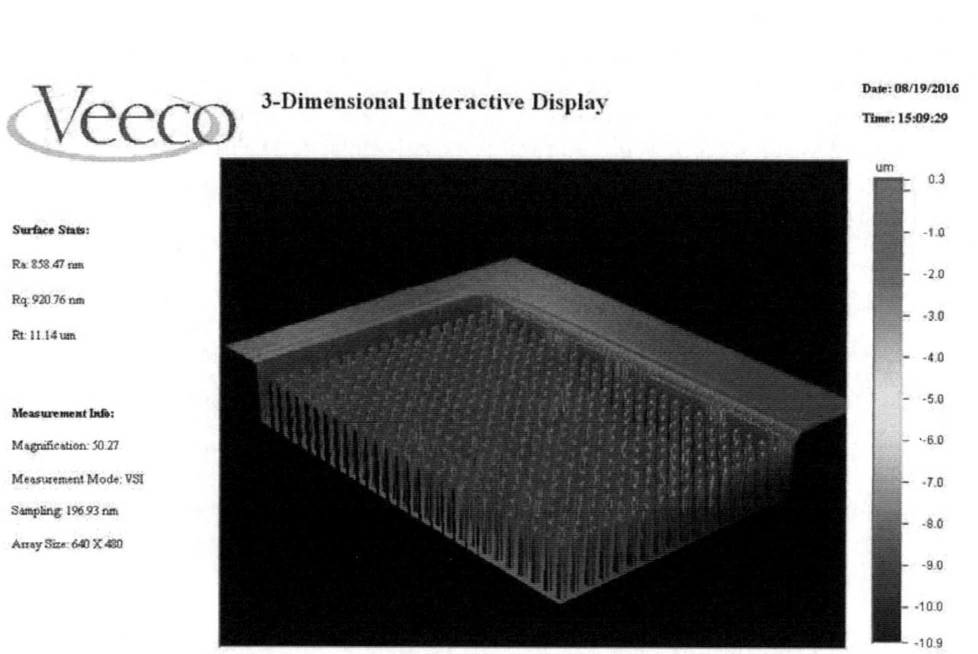

Figure 6. Wyko white light interferometer scan of the 2um microwire arrays which had a finished depth of 9-10um. The 3um versions had a finished depth of 13-14um.

6. Conclusions.

We have demonstrated the successful manufacture of microwire array targets with both 3um and 2um critical dimensions in 500um x 500um areas. Both types have now been used in a recent experimental campaign[4]. Further optimisation of the manufacturing processes could lead to improved targets of this type (e.g. narrower and/or taller microwires), should they be required for future experiments.

7. References.

1. Dimitri Khaghani et al. “Enhancing laser-driven proton acceleration by using micro-pillar arrays at high drive energy” Sci. Rep. 2017, 7: 11366.
2. Purvis MA, et al. Relativistic plasma nanophotonics for ultrahigh energy density physics. Nature Photonics. 2013;7:796–800. doi: 10.1038/nphoton.2013.217.
3. N. Wilke, A. Mulcahy, S.-R. Ye, A. Morrissey Tyndall “Process optimization and characterization of silicon microneedles fabricated by wet etch technology” Microelectronics Journal 36 (2005) 650–656.
4. "A bright laser-driven X-ray and particle source using microstructured silicon targets" submitted to Nature Physics.

Institute of Physics
Dirac House, Temple Back
Bristol BS1 6BE UK

ISSN: 1742-6588
ISBN 978-1-5108-7088-8

9 781510 870888

CURRAN ASSOCIATES INC.
proceedings.com